Navy RDT&E Planning in an Age of Transition

Navy RDT&E Planning in an Age of Transition

A Survey Guide to Contemporary Literature

Rodney P. Carlisle

A joint publication of the
Navy Laboratory/Center Coordinating Group and the
Naval Historical Center
Department of the Navy
Washington
1997

Library of Congress Cataloging-in-Publication Data

Carlisle, Rodney P.
Navy RDT&E planning in an age of transition : a survey guide to contemporary literature / Rodney P. Carlisle.
p. cm.
Includes bibliographical references and index.
ISBN 0–945274–37–8
1. Naval research—United States—History. I. Title.
V393.C38 1997
359'.07'0973—dc21 97–39907

Contents

Illustrations

Foreword

This survey guide is the second product to be published as the result of a collaborative effort of the Navy Laboratory/Center Coordinating Group (NLCCG) and the Naval Historical Center. The objective of this joint effort is to record Navy history associated with research, development, test, and evaluation (RDT&E) and the acquisition of Navy warfighting systems. The first of these publications, *Management of the U.S. Navy Research and Development Centers During the Cold War*, covered the 47-year period immediately following World War II and traced the reports of some eighty "blue ribbon" panels that examined operational and organizational issues confronting the Navy's in-house RDT&E shore structure. This report examines the various impacts of the rapidly changing international scene during the 1980s and 1990s on the Navy's policies and planning for its in-house technical community. It briefly looks at lessons that can be gained from a history of prior demobilizations during this century and draws parallels with the current drawdown being experienced by Defense. We hope that this guide will be beneficial to those who must plan and direct the Navy's transition to a sustained peacetime footing while assuring a modern, well-equipped Fleet poised to meet any and all potential threats to that peace which may appear on the horizon.

Dr. Rodney P. Carlisle, the author, is vice president of History Associates Incorporated and a frequent contributor to the body of literature on Navy RDT&E and its management. Among his recent books are *Powder and Propellants: Energetic Materials at Indian Head, Maryland, 1890–1990* and *Supplying the Nuclear Arsenal: American Production Reactors, 1942–1992*. Publication of *Where the Fleet Begins,* a comprehensive history of the David Taylor Research Center, now the Carderock Division of the Naval Surface Warfare Center, is scheduled for 1998.

A third publication, *The Relationship of Science and Technology: A Bibliographic Guide*, is also available. The views expressed herein are those of Dr. Carlisle, and not those of the Department of the Navy or any other agency of the U.S. Government.

K. K. Paige
Rear Admiral, U.S. Navy

Dr. William S. Dudley
Director of Naval History

Preface

This survey guide reviews the interplay between the changes in the international scene during the 1980s and 1990s and the Navy's policies towards its research, development, test, and evaluation (RDT&E) facilities. The emergence of a new set of world conditions and new national priorities during this period profoundly affected the structure of defense in general and, more gradually, began to affect the programming and shaping of research and development facilities. As the U.S. Navy readjusted RDT&E priorities in the late 1980s and early 1990s, it looked to the history of demobilization and conversion that followed each of the major wars in the twentieth century for suggestive parallels and lessons learned.

The guide has been prepared under contract with the Navy Laboratory/Center Coordinating Group (NLCCG) through History Associates Incorporated. For the most part primary documents cited here were available at the group's archives, formerly the Archives of the Director of Navy Laboratories, maintained at the White Oak Detachment of the Naval Surface Warfare Center. Those records were transferred to the Operational Archives of the Naval Historical Center in 1997. For the most part, documents are published reports readily available at the archives. Tapes of oral history interviews cited in this report are also at the Operational Archives. Cited secondary sources are widely available in academic libraries.

The term "Research and Development" is subject to much interpretation and debate. In the context of this report, the facilities discussed as part of the naval RDT&E establishment include the Naval Research Laboratory and the full-spectrum R&D engineering centers—designated as warfare centers in the 1990s. Other parts of the Navy's research establishment, such as oceanographic and medical research facilities, are not included in this study.

A chronology of events for the 1985–1993 period, showing the content of international developments and major actions taken and major documents issued by the Departments of Defense and the Navy over the same period, appears as an appendix to the guide.

The opinions expressed herein are mine alone and do not necessarily reflect those of the Departments of Defense and Navy.

Rodney P. Carlisle

1

Introduction and Background: Prior Demobilizations

In recent years, historians, journalists, and policy analysts have made comparisons between the transition to the post-Cold War period and earlier postwar periods and demobilization, drawing parallels and contrasts. This report surveys documents prepared by all three types of commentators and discusses the post-Cold War transition in defense procurement in the light of comparable prior transitions.

In regard to Research, Development, Test, and Evaluation (RDT&E) activities in the Navy, some experiences of World War I generated lessons applied in World War II. In turn, the role of RDT&E in World War II shaped a policy of continued emphasis on technological progress during post-World War II demobilization. That policy of maintaining a permanent RDT&E structure continued through both the Korean War and the Vietnam War and throughout the Cold War period. Over and over throughout the twentieth century, Defense and Navy Department planners sought to learn from the experience of recent history, with some degree of success.

Although policy planners repeatedly tried to draw lessons from recent events, research and development priorities during demobilization and interwar periods often resulted in poor preparedness for the crises which did develop. During its participation in World War I (1917–1918), World War II (1941–1945), Korea (1950–1953), and Vietnam (1964–1972), the United States frequently provided American soldiers, airmen, and seamen with marginal or outdated weapon systems because of a prior demobilization, drawdown, or delayed weapons development. The inadequate and outdated nature of the weapons was usually most apparent in the first battles or engagements of each war. As the war went on, the defense structure mobilized and the weapons improved: older, reliable weapons were produced in greater quantity, and partially developed weapons and systems were rushed into production. In each case the United States tried to rely upon its superior technology, hastily produced during the war, to substitute for massive manpower and to ameliorate American casualties.

Demobilization followed the end of each war, yet each demobilization was unique in scale and effect. Following the demobilization after World War II, a national debate led to the creation of a semipermanent defense industrial base and a large government-owned R&D establishment. The development of the Cold War, which lasted from 1946 until 1989, prevented full demobilization after the Korean War and the Vietnam War. Rather, as phases of the longer Cold War, each of these wars generated adjustments and lessons learned for naval technology and produced periods of shifting RDT&E emphasis and partial, uneven retrenchment in the procurement of new implements of war.

After a hiatus in 1947–1948, nuclear weapons and their delivery systems received continuous funding. Soviet advances in the late 1940s stimulated intense activity in developing nuclear warhead delivery systems, nuclear propulsion of submarines and ships, and improving antisubmarine technology. The "interwar" period 1953 to 1964, like the period 1918 to 1939, was one of concentrated development on *particular* systems. Meanwhile, funding the systems for delivering and countering conventional ordnance tended to be intermittent through the 1950s and 1960s. For such reasons, demobilization had sometimes contradictory and complex effects on naval RDT&E and on naval procurement policies in general. Furthermore, the patterns of lessons learned and applied from one demobilization period to the next were quite intricate.

World War I

U.S. Participation. Over the years 1914–1916, when the Great War was consuming Europe, the United States under President Woodrow Wilson steadfastly adhered to a neutral position, demonstrated and confirmed by a conscious policy *not* to build up a standing army; however, in 1916, the nation undertook a shipbuilding program to build a "Navy Second to None." Wilson worked to get the warring states to adhere to civilized codes of military engagement, to define their war goals, and to negotiate an armistice or peace. His intention was to use America's clearly announced neutrality policy as a bargaining tool with both the Allies and the Central Powers.

Although national policy under Wilson remained officially neutral, the United States expanded its seaborne trade with the British and French. That neutrality was not an "armed neutrality" but one of ostentatious lack of preparation. Ironically, it was the very low degree of U.S. preparedness that was a major element in Germany's calculation to open unrestricted submarine warfare against the ships of all nations, belligerent and neutrals alike, supplying Britain and France. That decision on 1 February 1917 forced the United States to respond and finally declare war on 7 April 1917.

The German High Command concluded that it would take at least a year for the United States to mobilize its industry to a war footing and to deploy effective numbers of troops in Europe. Within that year the German submarine blockade of Britain was expected to starve that nation into a negotiated peace. The German command had assumed that shipping would continue as it had, with fast ships attempting to outrun submarines and slower individual ships making tempting torpedo targets. But in early 1917, Admiral William Sims of the U.S. Navy, among others, urged the creation of a system of convoys, flanked by destroyers and armed auxiliaries such as converted trawlers, to protect cargo ships and tankers. Within a few months, the convoys foiled the German calculation of the amount of food and fuel supplies their submarines could interrupt, and Britain was not forced to bargain for peace before American troops could arrive in effective numbers.

However, the German estimate of a year's delay in U.S. troop mobilization was not far from being accurate. U.S. forces did not begin arriving in sufficient numbers to affect the course of the war until mid-1918, fully fourteen months after the United States had declared war. More than one million men were transported safely from the United States to Europe aboard converted fast passenger liners, none of which was attacked or damaged by submarines.[1]

During 1914–1917, when as a neutral the United States had not expanded its own forces, war orders by the British and French stretched parts of the U.S. defense industrial base. When the United States entered the war in April 1917, Atlas Powder Company and Hercules Powder Company, two successor firms created after a court-ordered breakup of the DuPont Company monopoly on smokeless powder, had already taken huge orders for the propellant from Britain and France; the companies could not meet sudden demands from the U.S. Army and Navy. As a consequence, the Navy expanded its powder factory at Indian Head, Maryland, doubling and redoubling its capacity over the nineteen months of U.S. participation in the war. Plans were laid for construction of hydroelectric-powered air-reduction facilities in Muscle Shoals, Alabama, to generate nitrates to fill the sudden shortage created by Allied purchase of natural nitrates from Chile.[2]

Delays in mobilization. In many areas of naval material supply, procurement came to fruition well after the Armistice in 1918. The first of the "flying boats," large seaplanes, for example, were built in the United States and flown across the Atlantic *three years after* the war ended. And these NC-9 craft were never built in the planned numbers. Merchant ships built of concrete were planned but not delivered, and Ford-built antisubmarine Eagle boats were delivered after the war. The Muscle Shoals facilities, planned to generate nitrates for the armaments program, were completed two years after the war, and their disposition became a political issue in the 1920s. In addition, the sudden completion of conventional-hull merchant ships in 1919 and 1920 led to a tanker and freighter glut, with disastrous economic impact in the 1920s.

International limitations on total tonnage of warships, set at the Washington Armament Conference of 1921–1922, compounded the over-supply problem for shipbuilders. The naval arms limitations led to scrapping of ships and cancellation of partially completed vessels, which in turn contributed to the long-range decline of the U.S. shipbuilding industry. In both areas of naval material supply—nitrate production and ship manufacture—the long start-up time not only delayed effective mobilization in war but left a major economic and political issue for the postwar generation in managing the facilities during peacetime.

Lessons learned for naval technology. Priorities for surface warfare weaponry were set by close examination of the major engagements during the war, and in this respect the war experience was a source for many specific "lessons learned." The Battle of Jutland from 31 May to 1 June 1916, the last engagement of main battle fleets, shaped British and, to an extent, United States R&D planning, particularly in surface warfare ordnance. Although the British succeeded in driving the German fleet back to port, they realized that the Germans had outgunned them. British armor-piercing shells tended to explode on contact, rather than penetrate the enemy's

U.S. Army aircraft under the command of General William "Billy" Mitchell conduct bombing tests on ex-German battleship *Osfriesland* off the Virginia capes.

NH 91090

armor. And British ships' magazines had proven vulnerable through their elevators to the gun rooms. German gunnery was more accurate, and use of starshells, smokescreens, and flashless powder had enhanced the German's ability to fight at night. The U.S. ordnance research agenda during the 1920s echoed the lessons of Jutland, as ordnance experts sought to improve fuzing and accuracy of fire, to develop illuminating shells, and to examine the location of, and protection around, ship magazines.

At Jutland, some of the artillery exchanges had approached nineteen thousand yards in range. As a consequence, the Navy's Bureau of Ordnance shifted all its range testing of naval guns from Indian Head to Dahlgren, Virginia, where a longer protected range over the Potomac River allowed for testing at those distances. Dahlgren remained a division of Indian Head until 1932.[3]

In July 1921 General William Mitchell conducted, under naval Bureau of Ordnance direction, the famous bombing tests off the Virginia capes against captured German vessels. Mitchell's flyers angered the naval officers conducting the tests, for they would not allow time between bombing runs for inspectors and engineers to survey ship damage, as had been agreed upon between the Army bombers and the naval experts. Mitchell argued that such delays would never occur in battle conditions and would hide the true effect of aircraft attack against ships. Naval officers grew bitter as Mitchell attempted to portray them as universally opposed to the development of air power. Yet, over the 1920s the Navy actively pursued the development of the aircraft carrier, developed and procured several dirigibles, and improved fighter aircraft, as well as the 5-inch, 25-caliber antiaircraft gun, recognizing the importance of an air arm to naval combat.

Secretary of the Navy Josephus Daniels had set up a civilian Naval Consulting Board during the war, which led to the establishment of the Naval Research Laboratory (NRL) in 1924. In the undersea area, work on torpedoes and mines continued at the Naval Torpedo Factory in Newport, Rhode Island, while study of underwater acoustics and radio communication for submarines went forward at the NRL. The Navy concentrated on developing the "Fleet Submarine," a long-ranging submarine capable of operating not only from main bases as a coastal defense weapons platform, but with the blue-water fleet as a commerce and warship raider. Work on these vessels required developing a strong diesel engineering program at the Engineering Experiment Station at Annapolis, Maryland, which flourished in the 1930s.

Armor-piercing shells, fewer duds in the production runs of shells, tracer shells for aircraft machine guns, improved safety buoys, gradual improvements in antiaircraft weapon accuracy, incandescent flare shells, and diesel engines with lower weight-to-horsepower ratios all proved to be valuable technologies in the next war. Perhaps even more important was the Navy's ability to maintain a small research capacity and a continuing staff at locations such as the laboratory at Newport, the Naval Research Laboratory in Washington, the Naval Powder Factory at Indian Head, the

Experimental Model Basin at the Washington Navy Yard, and the Engineering Experiment Station at Annapolis. While not lavishly funded, these "shore establishment" units limped along through the interwar years gradually improving upon a wide range of projects, devices, and systems.

In surface, air, and subsurface work, naval planners in the interwar years following 1920 had the foresight to keep research and development advancing. Although interwar development projects were quite mundane in contrast to the more spectacular weaponry developed in World War II, these ordinary and technically unsophisticated projects would prove to be crucial when the war came.

Interwar Developments Abroad

Work on relatively obscure, but essential, infrastructure and conventional equipment and weaponry indicates that RDT&E during the interwar years (1919–1939) is not a simple case of U.S. neglect but a more complicated picture of some advance and maintenance, coupled with avoiding investment in expensive new systems requiring breakthroughs in R&D. Japan, within warship tonnage limits set by the 1922 Washington Conference, developed advances in aircraft, torpedo, and ship technology that surprised the United States in war. In particular, their improvement of air-dropped torpedoes put them far ahead of U.S. technology in that system. The Mitsubishi-designed (and Nakijima-produced) *Zero* further surprised American pilots by consistently outclassing their own aircraft at the beginning of the conflict.[4]

A recent historical study of weapons-related research and development in several countries during the years 1918–1939, prepared for DOD's Office of Net Assessment (ONA) and entitled "Innovation in the Interwar Period," confirms that the picture of advance and retardation varied widely from nation to nation and in each case was quite complex.[5]

The ONA study argued that the differential development between the major powers was less a question of support or neglect than one of emphasis, *based on the cultural values and strategic views of the different nations*. In Germany, General Heinz Guderian developed strategies related to motorized units and tank warfare; the Luftwaffe developed a concept similar to what Americans called "close air support" in developing the dive-bomber. German naval technicians focused on construction of a few capital ships and on improvements to the submarine. As a continental power thinking forward to ground advance in Europe, Germany excelled in these areas of development while it lagged in other areas such as long-range aircraft and surface ships.

In the United States, aluminum air-frame development went forward, partly because of the rich civil airplane industry and growing long-distance private air transportation system. American geography was conducive to thinking about long-distance aircraft as an essential tool for strategic bombing. The ONA study noted that the U.S. Navy had learned the lessons of

World War I with regard to submarines and developed a vigorous training and development program around the long-distance submarine for interdiction of enemy shipping.

As an island nation concerned with maintaining a defensive posture, Britain developed radar detection equipment in the 1930s. The particularly unique British system established a widespread network to integrate air warning information, which proved of great value during the Battle of Britain. Study of the radar network established the groundwork for later British contributions to the field of operations research. British advances in sonar (called ASDIC) during the same period reflected concern over the vulnerability of vital shipborne supplies to interdiction by submarine and torpedo.[6]

World War II

Neutrality, 1939–1941. In the United States, some of the lessons of the wasted 1914–1916 period had been learned; the country once again enjoyed a period of neutrality during which it could build up its defenses. Although the political mood remained isolationist, President Franklin Roosevelt began a naval shipbuilding and modernization program under the

NH 91784

A 1940s view of the new David Taylor Model Basin, Carderock, Maryland.

National Industrial Recovery Act of 1933. World War II in Europe began on 1 September 1939. Over the next two years, the United States put its defenses in better order by establishing the Selective Service system in 1940, by completing and refurbishing old facilities and opening new ones, and by acquiring bases in British Caribbean colonies.

The Navy's Experimental Model Basin, built by Captain David Taylor in 1899 on the Anacostia River at the Washington Navy Yard, had been housed in a small warehouse-like brick building next to the river. The staff moved to new state-of-the-art facilities at Carderock in 1940, appropriately called the David Taylor Model Basin, where an accelerated program of ship design flourished with a focus on faster and more stable destroyers and new and larger aircraft carriers.[7]

Despite the greater degree of research and preparation in the pre-World War II years compared to the 1914–1916 period, on 7 December 1941 it is true that the U.S. fleet's ships and aircraft, weapons, and electronics were years—and in a few cases, decades—behind the technology already in the field and afloat by Germany, Japan, and Britain.

U.S. Technology in WWII. MIT president Vannevar Bush defined the problem in a letter to President Roosevelt in June 1940. Roosevelt responded by establishing the National Defense Research Committee and later expanding it with more powers as the Office of Scientific Research and Development (OSRD), under the leadership of Bush. The OSRD set the tone for wartime research and development and established "Big Science" in the United States. Research went forward on a host of new developments under government funding and supervision through contracts with academic institutions and private industry. OSRD efforts spawned advances in proximity fuzes, radar, Jet Assisted Take-Off (JATO), rockets, and the nuclear weapon itself. In contrast to the World War I period, however, the United States was able to field most of the major systems during the war years.

The most spectacular OSRD-sponsored project, the nuclear weapon, was turned over to the Army Corps of Engineers for development in 1942. Two billion dollars were expended to rush the development through over the next three years. The bomb was completed and ready in July 1945, just as the war in the Pacific was approaching the Japanese home islands.[8]

Prewar U.S. ignorance of Japanese advances in weaponry became legendary. For later generations the typical viewpoint held in 1941 by naval officers, War Department planners, and the American public towards Japanese technology and fighting ability is particularly instructive. During the 1930s American observers were well aware that Japan was militaristic, that it was supplying and arming territories in the mandated islands of the South Pacific, and that it was engaged in brutal aggression in China. Nevertheless, in mid-1941 Americans perceived Japan with some contempt, a non-European state which would be no match for a major western power. In terminology of the 1990s, Americans viewed Japan not as a world power but as a "Major Regional Power." It was inconceivable that an

Post-World War II research on the German V-2 rocket, such as the one pictured here at White Sands Proving Ground, New Mexico, led to the development of more effective missiles in the United States.

Asian nation with less than 15 percent of the GNP and with less than half the population of the United States could represent a threat. Thus, the successful air attacks on Pearl Harbor on 7 December and on British warships *Repulse* and *Prince of Wales* off Malaya on 10 December 1941 were not only a surprise but a rude awakening to Japanese superiority in weaponry and warfighting. Suddenly, a militaristic regional power had become a major threat to American and British national security. Americans pledged themselves to "remember Pearl Harbor," converting the sense of a lesson learned to a popular motto.

The performance of U.S. weapons in the first months of World War II is best described as "pitiful." Gordon Prange, in his definitive work on the Battle of Midway, showed how the material unpreparedness of both Army and Navy aircraft led to the loss of aircraft and aviators in the attack on the Japanese fleet. Although analysis of signal traffic had revealed Japanese plans, the surprise generated by superior U.S. intelligence was dissipated in the failed first raids on the Japanese task force of aircraft carriers and support ships. Prange provided a host of examples of flawed American technology: bombs that were duds, arming mechanisms that released bombs harmlessly over the ocean, guns that failed to fire, the useless handling of the F2A3 as a combat aircraft, and the mule-like performance of other aircraft. In general, when an airdropped U.S. torpedo did explode, it "was an event worth writing home about."[9]

Fortunately for the United States during the war years, Germany's decisions to underfund or delay research explained the failure of that nation to capitalize on its capacity to develop radar, jets, and rockets earlier, and to pursue development of the nuclear weapon. Hitler had decided in 1940 that the war would be short and ordered his researchers not to work on weaponry that would take more than six months to bring to the field. Despite these orders, General Walter Dornberger persisted in his work on the V-1 and V-2 rockets, leading to the successful employment of these weapons against targets in Britain and Holland. Dornberger recalled later that Hitler personally apologized for not believing earlier how important the work on the V-2 was.[10]

In contrast to American values with regard to Japan, American respect for German technical superiority was deep and pervasive. At the urging of Leo Szilard and Albert Einstein, President Roosevelt initiated the U.S. nuclear program precisely because published work on nuclear fission by Otto Hahn and Lisa Meitner in Germany suggested that Nazi Germany would develop the weapon. German acquisition of Czechoslovakia with its known deposits of uranium, together with a solid respect for German physics, engineering, and industrial capacity, convinced U.S. planners that they needed to develop the nuclear weapon to deter and possibly offset a parallel German development. The failure of Germany to take advantage of its lead in the nuclear field has been detailed in several thorough studies. From a naval perspective, it is interesting to note that German nuclear physicists placed more emphasis on developing a reactor as a source of power or ship-propulsion than on using nuclear fission to produce a weapon.[11]

2

The Cold War Era

Post World War II demobilization. Although the United States demobilized its servicemen, ordnance, aircraft, and naval fleet following World War II, the technological and scientific establishment created during the war was not entirely dismantled. As the Cold War began in 1946, scientists and policymakers worked to maintain and rebuild the research establishment, looking to the immediate past to demonstrate the urgency of their concerns.

Histories of the German technological effort written in the decade following World War II often carried the "moral," or lesson, for western policymakers that Nazi neglect of R&D had lost Germany the war. Several of those memoirs and histories became part of the domestic U.S. intellectual context in which Cold War policy was set. In particular, the Germans failed to develop microwave radar allowing Allied forces to cripple the Nazi U-boat fleet in 1942, and the late deployment of jet-propelled aircraft made little difference in Germany's command of the air.[12]

In the United States, scientists and science administrators took an active role in the postwar debate over the state of military R&D. In two books, *Science: the Endless Frontier* (1946) and *Modern Arms and Free Men* (1949), Vannevar Bush argued that modern science was essential to winning weapons technology. In order to fund the necessary research establishment, the government had to support science on a scale not possible without government aid in university or private laboratories. A scientific establishment and a development establishment had to be permanently maintained to keep the nation prepared for a determined adversary. With the emergence through the late 1940s of a U.S. consensus that the Soviet Union represented a military and political threat to western freedoms, Bush's warnings found a receptive audience. The war-built institutions such as the OSRD gave way to a network of government-owned laboratories, academic laboratories supported by federal contracts, and industrial defense contractors. Although several members of Congress argued for a centralized, government-controlled research establishment, what emerged was much more loosely organized and varied in structure. Nevertheless, the nation heeded the central message of Bush, James Conant, Karl Compton, Rear Admiral

Vannevar Bush, author of *Science: The Endless Frontier*, advocated large-scale federal support for scientific research and development, tying that support to America's need for a strong defense establishment in the postwar years.

Courtesy Karsh Studios and Carnegie Institution

Harold G. Bowen (head of the Navy's Office of Research and Inventions), and others that science and technological development had to be made part of the nation's defense posture.

In 1946 the Navy established the Office of Naval Research (ONR), headed by Admiral Bowen, as well as the Naval Research Advisory Committee (NRAC), which met at least annually to provide guidance and leadership to the research and development programs of the Navy. The best analysis of the origins of ONR shows how central that organization became to the progress of academic science in the immediate postwar years. Harvey Sapolsky, author of *Science and the Navy: The History of the Office of Naval Research*, concluded that ONR provided support to basic research from 1946 through 1950 relatively unnoticed by Navy line officers. ONR served as the principal federal agency supporting academic science prior to the establishment of the National Science Foundation (NSF), and NRAC continued to help develop policy and even choose personnel. In 1955, for example, NRAC helped select the research director for the Naval Research Laboratory. Under the administration of Secretary of Defense Charles E. Wilson during the Eisenhower administration, however, ONR's support for independent research was drawn down.[13]

Lessons learned for naval technology. In the immediate postwar years of the late 1940s, the Navy adapted to the nuclear age with several approaches. The Navy organized and participated actively in Operation Crossroads, the test of nuclear weapons' effects against ships at Bikini Atoll in the Marshall Islands in 1946. German advances in submarine propulsion, including the snorkel device that allowed undersea operation of diesels and the self-contained hydrogen-peroxide engine, led to experiments in the late 1940s at the Engineering Experiment Station with both snorkel systems and alternative engine systems. As early as 1947, submarine officers and researchers anticipated that the submarine of the future would be a true submarine vessel, rather than a submersible surface ship. As such, the new submarine would cruise and operate for long periods entirely without surfacing. Nuclear propulsion promised such a system that would be superior to the partial moves in that direction promised by snorkel operation and by self-contained innovative engine systems.

Sources for R&D direction. The research and development community of the Navy through the 1950s developed other institutions for policy planning besides NRAC. The contributions of several conferences in providing direction to submarine development have been analyzed by Gary Weir in a study of submarine technology, *Forged in War*. Weir shows how, in a series of conferences through the 1940s and 1950s, policy planners, scientists, and engineering officers met to set the direction of research. For example, in 1946 civilian scientists, led by Gaylord Harnwell of the University of California Division of War Research, successfully petitioned Admiral Bowen at ONR to request that the National Research Council establish a permanent Committee on Undersea Warfare (CUW). The CUW recommended research into subjects such as submarine self-noise, transport of sound undersea in the sound-fixing and ranging sound channel of the oceans (SOFAR), and the hydrodynamics of undersea vessels. These studies were inspired by captured advanced German submarine types developed at the end of World War II.

In early 1950, a committee chaired by Vice Admiral Francis Low began a comprehensive study of undersea warfare and recommended further research directions. Basing the investigation on a precise set of estimates regarding the size and quality of the Soviet submarine fleet, the Low group recommended over 100 initiatives and lines of future technical work in its report of 22 April 1950. The fact that the Soviets had confiscated several advanced German submarine types, including at least 13 equipped with snorkels, and several of Types XXI and XXIII, prompted the United States to move on several fronts. The Low group recommended pursuit of closed-cycle engines, capable of extended submerged operation; further work on nuclear propulsion; and development of a submarine capable of launching medium-range guided missiles. Even more perceptive than these specific plans, the Low Report commented on the need for further agenda-setting in technological areas "to further reorient our present Research and Development Program to counter these long-term threats." However, the report warned that the present R&D planning system failed to regularly assess the larger picture. The group suggested that a "Research and Development Review Board periodically assess the strategic and technological situation . . . as a factor pertinent to the intelligent assignment of priorities."[14]

Later that year, another group met at MIT, bringing together WWII experts from the fields of nuclear research, radar, sonar, missiles, fire control, and proximity fuses among others. This "Hartwell Project" resulted in recommendations for work in a wide variety of areas, including hunter-killer submarines, radar, enhancement of long-range detection and low-frequency sonar, nuclear depth charges, and other submarine countermeasures and detection methods. Between the Low Report and the Hartwell Project, the Navy science community set priorities in submarine and antisubmarine technology through the early 1950s.[15]

Soviet success in developing an atomic bomb in 1949 and a hydrogen bomb in 1953 spurred on the U.S. nuclear program, but some areas of "conventional" RDT&E did not fare as well during the same period. Even with the burst of war orders for Korea, the older bureau-owned and -operated facilities tended to be neglected in an era which favored private sector enterprise over government ownership, and nuclear propulsion over traditional systems. Between the end of World War II and about 1956, both secondary literature on naval weapons and primary documentation from facilities like David Taylor Model Basin and the Engineering Experiment Station at Annapolis indicate a period of very selected focus in research activity. Work on JATOs, snorkel and alternate propulsion systems at the Engineering Experiment Station, and some wind-tunnel work on jet aircraft at David Taylor stood out as areas of activity during an extended period of cutbacks, deferred maintenance, and institutional survival. The studies by CUW, the Low group, and the Hartwell group helped give policy direction and a focus to research through the late 1940s and 1950s.

What the studies by Sapolsky and Weir suggest is that even in a period of national demobilization and restricted budget for naval research, intelligent planning and concerted focus on the implications of the future threat could provide a stimulus to research which would have long-range benefits and develop pressures within the naval research community for particular types of R&D. Even as shipyards closed and budgets decreased, and before a national recognition of the threat imposed by Soviet technical advances had become widespread, the efforts of Bowen, and the Hartwell and Low groups kept American naval research focused on a number of important long-range goals.

Yet the focus that emerged from the Hartwell and Low conferences was on what a later generation would call high technology. The U.S. Navy engaged in separate arms races with the Soviet navy in nuclear propulsion, nuclear weaponry, underwater acoustics, and missile development. However, the actual combat of the Cold War never pitted the United States against the Soviet Union. In Korea, Vietnam, and later in places such as Nicaragua, Angola, and Afghanistan, both nations worked through surrogate powers, through forces that relied heavily on traditional conventional weaponry and sometimes on quite primitive guerrilla techniques. Most naval high-technology weapons were never used against each other, but instead the two superpowers engaged the "conventional" or low-technology weapons of the surrogate countries. This factor made the lessons of mobilization and demobilization even more complex.

Korean War: Lessons learned for naval technology. During the Korean War, 1950–1953, the Navy learned a number of specific lessons regarding the lack of material preparedness. In a thorough and perceptive study of the history of mine countermeasures (MCM), *Damn the Torpedoes*, Tamara E. Melia demonstrated that some forgotten lessons from World War II were relearned during Korea.

The successful operational MCM efforts in World War II limited MCM development after the war. Despite rapid advances in influence mine development, the largest mine threat during the war remained the antique, World War I moored contact mine. To a budget-conscious Navy, wartime dependence by foreign navies on this cheap, easily countered mine made continued funding for developing defenses against more modern influence mines all the more unnecessary. Proponents of MCM readiness had trouble making themselves heard in peacetime.[16]

As General Douglas MacArthur advanced on North Korea, naval forces found more than three thousand mines in a 400-square-mile field at Wonsan Harbor, effectively stopping a coordinated sea assault to coincide with the land advance on Pyongyang. In effect, fifty thousand men and 250 ships of the U.S. Navy were held at bay for over a week by an assortment of influence and contact mines. The North Korean "antique" mines pointed out a "critical hole" in U.S. naval preparedness.[17]

Naval R&D in the Post-Korea Period

Admiral Arleigh Burke was appointed Chief of Naval Operations (CNO) in August 1955, and shortly after taking office, he requested that the CUW undertake a study of antisubmarine warfare. In 1956, in response to this request, and in the light of nuclear submarine technology and the apparent ability of the Soviet Union to emulate and keep up in technology, the CUW proposed a summer study under the direction of Columbus Iselin, director of the Woods Hole Oceanographic Institution in Massachusetts. "Project Nobska," as the conference was called, brought together a wide range of civilian and naval scientists. The group generated an extensive report, "The Implications of Advanced Design on Undersea Warfare." Working from similar premises as the smaller Low group did in 1950, the Nobska conferees looked at separate specific technical problems in more depth. As a whole, the group advocated a broad spectrum of technical and scientific and engineering projects.

Altogether, the Nobska conferees split into eight technical groups and six systems groups, each offering recommendations for future research. The technical groups studied (1) weapons effects and limitations, (2) propulsion, (3) navigation and communication, (4) detection, (5) missiles and torpedoes, (6) operational analysis, (7) ships for undersea warfare, and (8) human factors in undersea warfare. The systems groups considered (1) pro-

NP/45H 36947 Courtesy Naval Weapons Center, China Lake, CA

Captain Levering Smith, shown here at the Naval Ordnance Test Station, China Lake, in the early 1950s when he served as associate technical director. He became head of the Special Projects Technical Division designing the first ballistic missile submarine.

tection of ships at sea, (2) strategic uses of the undersea, (3) containment of submarines and defense of the continental United States, (4) overseas transport vehicles in undersea warfare, (5) overall concepts and costs, and (6) implications of political trends on undersea warfare.

Edward Teller, who served on the weapons effect group with others from the nuclear community, including I. I. Rabi and Harold Agnew, provided one of the key elements in the eventual success of the Polaris submarine concept by suggesting that missiles could be tailored to submarine requirements. A number of recommendations from the Nobska group shaped the direction of future research on ship machinery noise quieting, improved seaworthiness for antisubmarine surface ships, study of oceanography, and deep sea mines.[18]

The core of the Nobska recommendations on advanced submarine work reflected how the launching of *Nautilus* in 1954 had vividly demonstrated new possibilities in the nature of undersea warfare. Admiral Hyman Rickover's highly personal management approach was modified, expanded,

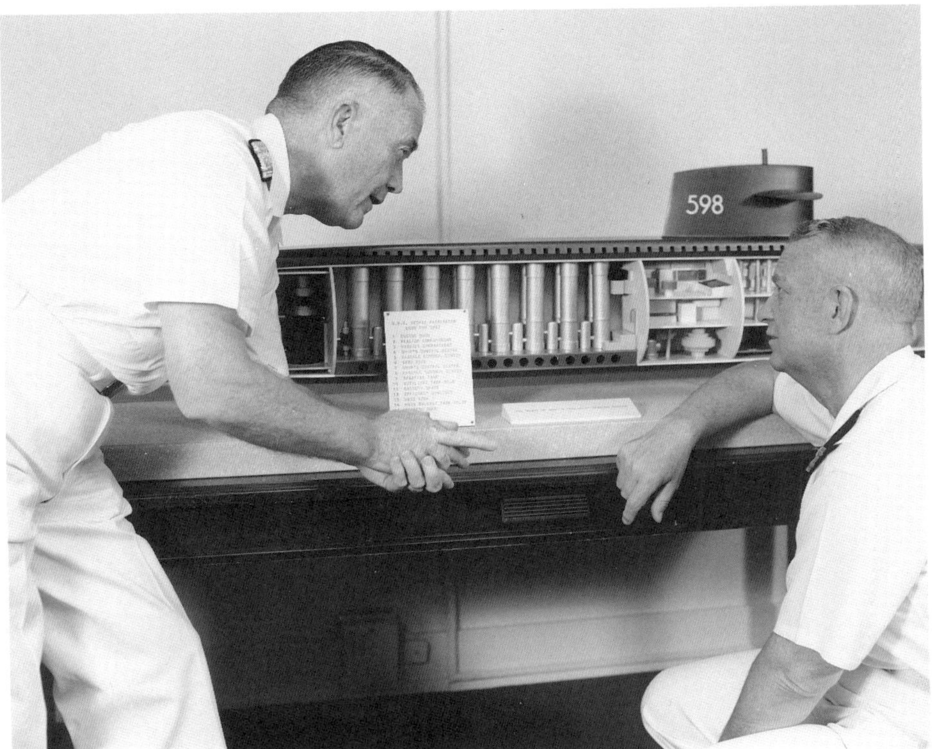

Rear Admiral William Raborn, head of the Special Projects Office for Polaris, and Chief of Naval Operations Arleigh Burke examine a model of *George Washington* (SSBN 598), the first fleet ballistic missile submarine.

and developed further with the Polaris program under Rear Admirals William F. Raborn and Levering Smith, which began in the late 1950s, to develop a system for underwater launching of intermediate range ballistic missiles from nuclear-propelled submarines. Like the *Nautilus*, the Polaris missile system and its successors were developed through Special Project Offices rather than through the traditional bureau-managed facilities of the naval shore establishment. The excellent historical and management studies of the submarine and the Polaris systems reveal that these two areas of platform and weapons development flourished while more conventional weapon systems languished in the early to mid-1950s. For these nuclear-related systems, at least, the early Cold War period was the equivalent of a long-term mobilization.

Following the Korean War, the Navy gained a nuclear strike capability. Nuclear weapons included not only Submarine-Launched Ballistic Missiles (SLBMs) and "standoff" ship and air-launched missiles but a variety of antisubmarine nuclear devices. Although the warheads themselves were designed and manufactured in Atomic Energy Commission (AEC) facilities,

President John F. Kennedy, a proponent of expanded funding for science, watches the launching of a Polaris missile from *Andrew Jackson* (SSBN 619).

L-file

the delivery systems actively engaged Navy shore establishment personnel and programs at Naval Ordnance Laboratory, White Oak, and Naval Ordnance Test Station, China Lake.

The SLBM program began in the summer of 1956 as a response to several studies including Project Nobska, which recommended an SLBM capability to be deployed by 1965. Anticipating that nuclear weapons could be reduced in weight by the time the project was completed, the Naval Ordnance Laboratory designed an arming and fuzing system to be integrated with a projected lightweight warhead designed by the Sandia and University of California Radiation Laboratories. With the successful Soviet launching of Sputnik in August 1957, the Polaris program was accelerated to be deployable by 1960 with a 1,200-mile range.

Polaris propellants were developed at Allegheny Ballistics Laboratory and tested at Naval Ordnance Station, Indian Head, Maryland. Manufacture of the propellant grains went forward at Indian Head in 1960–1965; production was later taken over by the private firm Hercules at its Bacchus plant. Meanwhile, the development of the submarine went forward, with the first Polaris submarine, *George Washington*, commissioned in December 1959. The submarine began its first operational patrol, carrying sixteen missiles armed with W-47 nuclear warheads, in November 1960. By

LHL L077146 Courtesy Naval Weapons Center, China Lake, CA

An ASROC ready for a crosswind firing test at Naval Ordnance Test Station, China Lake, CA, 1957.

1965 eighteen Polaris submarines had joined the fleet, with a total of 288 armed launch tubes.

During the 1960s the Navy developed the next generation of missiles, Poseidon, designed to carry multiple independently targeted reentry vehicles (MIRVs). The first Poseidon-class submarine went on patrol in 1971; by 1975, twenty-seven Poseidon submarines were in the fleet. Each missile on board the submarines carried ten warheads. Later, the Poseidon missiles were replaced with Trident I missiles, which carried fewer warheads but had longer range than the Poseidon.

During the 1950s, the Navy also focused on antisubmarine research as part of the response to the changing nature of the submarine. In October 1951 the Naval Ordnance Laboratory at White Oak reported on the feasibility of a nuclear depth charge as an antisubmarine weapon. Designed by the AEC and installed in a casing developed at White Oak, the Alias Betty was the first of these weapons added to the fleet. First deployed in 1955, the weapon was air-dropped with a retarding parachute; more than two hundred were produced, and all were retired by 1960.

KN-487

Light guided missile cruiser *Galveston* (CLG 3) is underway for the first at-sea firing of a Talos missile, February 1959.

In the mid-1950s the Navy and the AEC began discussing an atomic warhead for the Anti-Submarine Torpedo Ordnance Rocket (ASTOR). Although ASTOR was deployed between 1959 and 1962, the Navy discontinued its use since it required a noisy active sonar targeting system. In the early 1960s, as the submarine fleet was moving to a silencing program following Nobska recommendations, the promising development of a rocket-launched torpedo was underway. Originally designed around a JATO rocket on a Talos booster, the system became known as the Anti-Submarine Rocket (ASROC). Developmental work on ASROC at China Lake went forward while the AEC designed a low-yield warhead. The first ASROCs were produced for deployment in the fleet in 1961, and retirement of the system did not begin until 1974. In addition, the submarine-launched anti-submarine torpedo with a nuclear warhead, or SUBROC system, was designed. Planning for that system began in 1957, and it went into production in 1964.[19]

The Navy moved rapidly in the area of rocket propulsion for conventional ordnance as well as nuclear warheads in the late 1950s and early 1960s. In addition to Talos, ASROC, and SUBROC, China Lake and Indian

Head were actively developing and testing at least ten other solid-fuel propelled rocket systems, including Bullpup, Shrike, Sidewinder, Sparrow, Tartar, Terrier, and Zuni. The period between the end of the Korean War and the beginning of the Vietnam War was hardly a typical "demobilization." Rather, that decade saw hectic weapons development in the face of the Soviet capability demonstrated not only by Sputnik but by successful ICBM launchings and hydrogen-bomb testing.

The recommendations of the prestigious group assembled for Project Nobska in 1956 appeared to have shaped the agenda for much of Navy R&D over the decade that followed. In the midst of the Cold War, and in an "interwar" period following the end of the Korean War, the long-term thinking of this group, based upon an analysis of contemporary scientific, technical, and international political groups, clearly gave added impetus to crucial developments in undersea warfare.

In aircraft development, the Navy's "last gunfighter," the Chance-Vought F-8 Crusader, first deployed in 1957, continued deployment during this same period. Designed as a fighter to match the Soviet MIG 17 and MIG 19, the single-seater fighter aircraft served as an interceptor against Soviet Bear and Bison bombers. The F-8s, with superior speed and maneuverability, scrambled to meet the Soviet aircraft in several Cold War tests of nerves. In 1962, after the downing of a high-altitude U-2 reconnaissance aircraft over Cuba, Marine and Navy F-8s served as photo-reconnaissance craft, logging in thousands of pictures of installations there during the October missile crisis.

An effort to increase R&D in surface and air ordnance followed the relatively short Korean War. A postwar study conducted by Vice Admiral Thomas Connolly, Deputy Chief of Naval Operations (Air) and a former experimental officer at China Lake, brought attention to the aging quality of nonnuclear ordnance. Although it had been possible to conduct the Korean War using World War II vintage weapons, by the late 1950s, these stocks were depleted and the weaponry was becoming obsolete. As a consequence of the Connolly study and the realization of increasing obsolescence, the Navy started a program to develop new conventional weapons at the end of the Eisenhower administration. An example is the work that began in 1959 at China Lake on the later famous television-guided glide bomb, the Walleye. Again, a study gave focus to the issue and resulted in targeted research.

Cold War technological competition with USSR. The Soviet launching of an earth-orbiting satellite in 1957 proved to the American public that the Soviet Union had developed the capability to launch intercontinental missiles. In response to Sputnik, federal funding for U.S. science began to move forward at a greater rate. Funding and expansion under Presidents John Kennedy and Lyndon Johnson in the early and mid-1960s revitalized the government's R&D facilities. Under the leadership of Defense Secretary Robert McNamara, the Navy integrated RDT&E into the procurement of new systems. Meanwhile, both advocates and opponents of high-level defense expenditures had adapted President

Eisenhower's phrase, referring to the interlocking network of defense department offices and industrial suppliers as the "Military-Industrial Complex."

Nevertheless, during the early 1960s, several articulate spokesmen established the rationale for relying on an in-house capacity for R&D rather than an external, privately owned or industrially based research capability. The technical director at China Lake, William B. McClean, stated at a 1961 Symposium:

> I believe that the government must do its own research within its Civil Service Laboratories so that it will have the ideas, the competence, and the capability to say in what directions the work should proceed and what objectives it should achieve in those areas where the government has the sole responsibility, such as military research and development.[20]

In "Report to the President on Government Contracting for Research and Development" (Bell Report) in May 1962, David Bell noted:

> We regard it as axiomatic that policy decisions respecting the Government's research and development programs . . . must be made by full-time Government officials clearly responsible to the President and to the Congress. . . . These are basic functions of management which cannot be transferred to any contractor if we are to have proper accountability for the performance of public function and for the use of public funds.

A number of policy statements issued from the 1960s through 1980 stressed the need to have an in-house R&D capability in order to be a "smart buyer." Perhaps the clearest articulation of this position came from William J. Perry, then Under Secretary of Defense, in 1980:

> The function of selecting among technical alternatives requires internal technical capability of sufficient breadth, depth and continuity to assure that the public interest is served.[21]

In aircraft, nuclear weaponry, submarine development, and missiles, the decade after the Korean War saw rapid progress in naval R&D and in deployed systems. Most of the systems were designed in response to specific aspects of the Soviet threat; when the Navy engaged in Vietnam, some of those systems could be adapted. Nevertheless, the Navy's arsenal had to be rapidly converted to the specific reality represented by Vietnam.

Vietnam: technological catch-up. Despite the invigorated R&D environment in the early 1960s and a rationally planned program of strengthening naval RDT&E facilities, Navy preparedness for war in Vietnam was in a state of partial neglect when the Gulf of Tonkin resolution rapidly escalated the war in 1964. Ed Marolda has demonstrated the nature of that neglect in his definitive history of the Navy in Vietnam. Continued concentration on nuclear ordnance and nuclear weapons delivery capacity between 1953 and 1964 had led to the failure to purchase even sufficient conventional weaponry, let alone to move rapidly into new development for such systems. In effect, despite intensive preparation during the Kennedy and Johnson administrations, the Navy still operated at "mar-

LHL 154380 Courtesy Naval Weapons Center, China Lake, CA

An A-4 Skyhawk attack plane launches a Walleye television-guided bomb over China Lake, CA, October 1969.

ginal levels" in its state of readiness when the Vietnam war escalated. Marolda's study confirms the observation that nuclear and missile advances had somewhat overshadowed the importance of conventional ordnance. Such ordinary systems as mine countermeasures, artillery, and conventional ship design had languished.[22]

Even so, several improved systems developed during the late 1950s and early 1960s entered production and proved very useful during the Vietnam War. The Walleye TV-guided glide bomb, under development since 1959, was deployed in 1967. It represented the first air-launched guided air-to-surface weapon ever used in combat over an extended period of time. The Snakeye was another China Lake development between the Korean War and the Vietnam War. Snakeye was not a missile but rather a system of fins attached to conventional bombs to retard the bomb trajectory sufficiently to allow the aircraft to clear the area safely. Work on the Snakeye began in 1960, and regular production began in 1964.[23]

Essentially an "interwar" fighter aircraft designed to oppose the MIG, the F-8 Crusader saw action in a variety of roles early in the Vietnam conflict before it was replaced with the two-seater Phantom in 1967. In Vietnam, the aircraft found an early role as a "MIG-Killer," downing some nineteen enemy aircraft, mostly using Sidewinder missiles rather than gunfire. North Vietnamese radar-guided missiles and flak deployed against massive high-altitude bombers in Operation Rolling Thunder required a ground-suppression aircraft; the F-8 played that role, employing Zuni rockets aimed against North Vietnamese ground antiaircraft emplacements. The 1950s-generation F-8 served for three full years in Vietnam before giving way to later designs.

As in former wars, developers found themselves under intense pressure to put in place hardware that had been in the conceptual or design stages for years, or to respond to technological threats emerging in the enemy's arsenal. In 1968, for example, the Naval Weapons Center at China Lake developed the first active-optical fuses using solid-state lasers. The fuse, developed in ninety days and designated DSU-10, was a proximity sensor immune to electronic countermeasures and was used for the Standard Anti-Radiation Missile. Altogether, some fifty Vietnam War-specific tasks were addressed at the China Lake facility alone. Other developments in the 1968–1970 period included a ground beacon for use by ground troops to identify themselves to friendly attack aircraft, map illuminators, and a lightweight gun pod for the Marines.[24]

Other parts of the naval procurement establishment responded similarly during the Vietnam War. At Indian Head, Maryland, for example, the Naval Ordnance Station produced a host of new products (many developed at China Lake) during the period 1966–1972, including an improved Zuni rocket fuel with better specific impulse, rocket-assisted projectiles, and long-range bombardment ammunition. Nevertheless, one of the mainstay missiles of the naval air arm was the 2.75-inch "Mighty Mouse," or Folding Fin Aircraft Rocket (FFAR). This World War II weapon had actually been out of production between 1957 and 1965 so its production was restarted. Originally designed to be an air-to-air weapon, the rocket had emerged in the Korean War as the air-to-ground standard, and the Navy had to gear up production at the Naval Ordnance Station with assembly at McAlester, Oklahoma.

Again, in the area of mine countermeasures, the Navy as a whole was reminded during the Vietnam conflict of the lessons of World War II and Korea, although specialists at the Navy Mine Defense Laboratory in Panama City, Florida, had never forgotten those lessons.[25] In Vietnam the Navy developed a shallow water capability, a so-called "brown-water" fleet of minesweepers. However, after the war, this particular technology was once again neglected.

In the area of air-launched ordnance, the hiatus in production and development of the 2.75-inch rocket led to the revival of a World War II design for use in Vietnam. The missile used in 1970 was little different from the cellulose-wrapped extruded solid-fuel rocket developed at the California Institute of Technology in 1943–1944. Literally produced by the millions, the 2.75-inch rocket over the years outranked every type of ammunition or ordnance in quantity fired except for small-arms bullets.[26]

Naval technology following Vietnam. Despite the increase in war-specific missions during the Vietnam War, a noticeable decline in research programs and new projects began to set in with retrenchment by 1971.[27] Following Vietnam, the realization of the need for contemporary procurement led to further reforms of the material side of Navy. By the mid- to late 1970s, the United States had secured a lead in nuclear submarines, super carriers, air-launched guided missiles, and air-delivered ordnance. Later, the Navy held a technology lead over potential adversaries in ship-launched missiles, air-cushion landing craft, and several other naval weapons and ship systems.[28]

A few of the technical leads in nonnuclear areas came from quick-response programs generated during the Vietnam War. Among the devices developed in that period which entered the naval arsenal were the Anti-Radiation Missile (ARM) series, the Rockeye II cluster bomb, forward-looking infrared (FLIR) sensors used in the Night Observation Gunship, the Sea Chaparral air-defense rocket, and the chaff decoy rocket (Chaffroc) system to interfere with enemy radar. Some of the rapidly developed systems of the Vietnam era were the basis for systems employed nearly twenty years later in the Persian Gulf.

Cycles of "Lessons Learned"

Tamara Melia's thesis regarding the failure to remember lessons learned in mine countermeasures can be applied more broadly to other technical systems. Melia examined questions of developing an operational force and a cadre of trained and experienced officers, as well as the technology, and of making MCM a viable career path for officers. The waves of "lessons learned" and then forgotten in this one area of technology are quite typical of the whole issue of naval material preparedness in a number of nonnuclear fields.

Each war showed the need for keeping up R&D in time of peace; yet, following each war, a gradual, and sometimes a sudden, decline in R&D occurred. The concept that R&D should continue in peacetime seemed to be remembered from World War I to World War II and then applied in the late 1940s, particularly with the organization of the Office of Scientific Research and Development (OSRD) and later the establishment of the Office of Naval Research in 1946, modeled on OSRD. Even though R&D and Big Science received a push in the post-World War II era for the nuclear field, conventional areas of technology were again neglected, and some fell by the wayside. In effect, some of the more glamorous projects like nuclear weapons, nuclear-propelled submarines, and improved missiles received attention, while the more ordinary systems such as mine countermeasures, air-to-air extruded rockets, and engine systems were quite neglected. In such areas, a few well-proven technologies stayed in the inventory, remaining little changed for decades. Careful study and promotion of focused research in the areas of submarine and antisubmarine warfare, quieting of propellers and engines, nuclear propulsion and nuclear weapons, and missiles had generated concrete advances in those areas. Meanwhile, conventional weaponry in the areas of mines and ordnance, which would be essential in the wars actually fought in Korea and Vietnam, received much less attention, leading to a degree of unpreparedness in both of those wars.

The civilian-operated AEC cooperated with the services in developing nuclear weapons that could be delivered from a variety of air and sea platforms. Concerns with the Soviet lead in submarines led to a concerted effort that generated first *Nautilus* and then Polaris. Similarly, shocked by Soviet progress in nuclear weaponry and missile technology in the 1950s, the United States launched into an intense period of rocket and nuclear development in which the Navy actively participated well into the early 1970s.

With regard to the nuclear weapon and its delivery systems, the United States appeared to heed Vannevar Bush's warning that only with "modern arms" could "free men" remain free. Nevertheless, in areas of mundane conventional weaponry, such as mines and small arms, the arsenals of minor regional powers like North Korea and North Vietnam proved formidable indeed. This lesson, which World War II had taught to the specialists at Panama City, seemed not to be recognized by Congress until U.S. troops and seamen were actively engaged against those conventional systems. The fact that in World War II a "Major Regional Power," that is, Japan, could challenge and have initial successes against the United States remains a pertinent lesson of that period as the world of the 1990s realigns around just such powers in the Middle East, in South Asia, and elsewhere.

3

An Era of Transitions

After Vietnam, many political leaders expressed fears that excessive emphasis on defense priorities and neglect of pressing social and infrastructure problems had left the nation's basic health and poverty problems unaddressed. With an increasing skepticism regarding the nature of the Soviet threat, a widening political constituency during the 1980s called for full conversion from a high state of military readiness to one of peacetime priorities. With the political and economic collapse of the Soviet Union, the dissolution of the Warsaw Pact, and the elimination of Communist Party control of all but a few nations, the Cold War was declared over by 1989. From then on an immediate "peace dividend" was expected in the form of reduced defense expenditures, which might free funds for other pressing social needs.

As downsizing struck the services, both the naval RDT&E community and the prospective fleet were targeted for reductions. The challenge for the 1990s would be to maintain a sufficient and appropriate size fleet and naval air arm to protect national security in a world of changing alignments and, at the same time, keep continuity of required research and development and industrial strength. As the public debated the economic impact of the reduction, a number of internal Department of Defense and Navy advisory reports advocated carefully focusing resources on RDT&E and shaping more cost-effective methods of procuring new systems.

In the world of the late twentieth century and the beginning of the twenty-first century, it is clear that the United States, if it wishes to put forward a foreign policy in accord with its economic strength and its vision of human rights, must have a military that has the highest technological capabilities. Although a national consensus emerged between 1988 and 1995 that defense budgets should be drastically cut, advisors to both the Bush and Clinton administrations recognized that the remaining defense budget would have to be carefully managed and directed, and that wasteful procedures would have to be eliminated. The military would have to move towards a modern and constantly modernizing force sufficient to meet forthcoming challenges. On the other hand, advocates of long-deferred social and economic programs believed that the end of the Cold War presented an opportunity for diverting resources away from defense to domestic issues, thus abandoning the Vannevar Bush concept that a technologically advanced defense was essential to U.S. foreign policy goals. Some of the sophisticated advice on how to best maintain defense technical superiority appeared to disappear, at least temporarily, in the chaotic world of defense policy making and priority setting in Washington.

Overview: A Period of Multiple Transitions

In the period 1985–1993, the nature of international relations underwent momentous changes. At first perceived as a "thaw" in the Cold War, changes within the USSR led to a series of arms control agreements, the withdrawal of Soviet troops from Warsaw Pact states, the lessening of international tensions between the United States and the Soviet Bloc, and, finally, the dissolution of the Soviet Union itself by 1992.

In the United States these changes stunned Presidents, members of Congress, and policymakers for defense and foreign affairs. The changes came so rapidly, and they were so fundamental, that old doctrines persisted for years. But by the early 1990s, a cacophony of voices suggested policy adjustments in contradictory directions. For administrators in the Navy's community of research and development centers, the period was one of chaos and change. Decisions taken in the mid- to late 1980s had stripped the community of its central leadership. The Navy's RDT&E community faced the changes without the benefit of a single voice or coordinating office to develop policy or to convert the directives of the President, the Secretary of Defense, or the Joint Chiefs of Staff into specific action. No conferences or advisory groups with the scientific prestige of the Nobska group were convened. Over the late 1980s and early 1990s, the Navy sought to address the leadership and coordination gap, at first through a single coordinating systems command, the Space and Naval Warfare Systems Command (SPAWAR), discussed in section 4, and later through linked coordinating committees, discussed in section 8.

Despite indications of weakening Soviet power through the mid-1980s, it was only in November 1988 that the first western head of government acknowledged the change. In that month, Prime Minister Margaret Thatcher took the lead in pointing out that Soviet troop reductions in Eastern Europe meant the "end of the Cold War."[29] By 1989, many editorialists and policy analysts in the United States spoke of entering a "post-Cold War" era.

The first concerted response in the United States to recognizing the change was to cut defense spending. Members of Congress and journalists anticipated in 1989 that reduced international tensions could lead to a "peace dividend" in the sense of reduced federal expenditures on defense. Defense budgets and actual defense expenditures for fiscal years 1989 through 1993 showed a decline, as reflected in the following figures, not adjusted for inflation.[30]

Defense Outlay

Fiscal Year	Dollars in Billions	Percent of Federal Outlay
1981	167.6	23.2
1982	185.3	24.6
1983	205.9	26.0
1984	227.4	26.7
1985	262.7	26.7
1986	273.3	27.8
1987	281.9	28.1
1988	290.3	27.3
1989*	303.6	26.6
1990	299.3	23.9
1991	273.2	20.6
1992	296.3	21.6
1993	290.6	19.6

Dollar amounts include Department of Energy nuclear expenses and are in "then-year" dollars, unadjusted for inflation. Outlays as percentage of expenditures reflect the effect of inflation as well as increases and decreases in direct defense expenses. Figures are expended, not budgeted, amounts.

Source: Office of Management and Budget, *Historical Data and Alternatives for the Future*, January 1993.

*The year "Peace Dividend" discussions began

While the dollar amount expended peaked in 1989, the percentage of the federal budget expended on defense had peaked two years earlier in 1987. Both expended amounts and percentages generally declined after 1990, although the 1992 outlay went up, reflecting the replacement of stores and equipment following Operation Desert Storm. The Office of Technology Assessment (OTA) prepared a report in February 1992 that detailed the national economic impact of reduced defense expenditures and placed the end of the Cold War in 1991, not 1989–1990. Looking at the budget figures, however, it is clear that the perceived end of the Cold War began to have an effect on defense expenditures rather promptly.[31]

In the mid-1980s, several documents and pieces of legislation that had set out the goals for the Defense Department under the Reagan administration bore on Navy R&D issues: the Packard Commission Report, the Goldwater-Nichols Reorganization Act, and several less well-known documents stressing privatization, technology transfer, and a reduction of the bureaucracy. These documents, issued in the last years of the Cold War, continued to set the baseline of R&D policy through the beginning of the transition years. Some of the policies in these documents grew increasingly irrelevant as events raced ahead; other policies could be adapted to the new reality.

In 1986 President Ronald Reagan ordered a study of defense management by a specially appointed commission. The resulting recommendations, conceived against the background of the 1970s, continued to shape R&D

David Packard, co-founder of the Hewlett-Packard Company, served as chairman of a commission appointed by President Ronald Reagan to review defense management.

Courtesy Hewlett-Packard Company

management in the DOD into the late 1980s as the Cold War ended. The "Report of the White House Science Council, Federal Laboratory Review Panel," or "Packard Commission Report," was named after the group's chair, David Packard, chairman of the board of the Hewlett-Packard Company, who had served as Deputy Secretary of Defense under President Richard Nixon. The Packard Commission Report emphasized a series of ways in which the defense laboratories could benefit from practices common in the private sector.

In summary, the Packard Commission recommended:

- program stability or multiyear funding[32]
- clear command channels
- management by exception, that is, reporting which focused on deviations from agreed baseline[33]
- smaller staffs devoted to management
- closer communication between producers and users of research and development
- lower cost approaches using commercial products and commercial-style competition[34]

The Packard Commission Report continued to represent guidelines for DOD over the next few years, and later reports on defense management would refer to this document as a guide to improvement. In particular, the goal of cost reduction through greater reliance on commercial products, the call for reducing the number of management personnel, and the closer responsiveness of research producers to product users would all have continuing appeal when the President and Congress attempted to reduce the size of the federal bureaucracy and, later, to reduce its budget.[35]

In 1986 President Ronald Reagan signed the Goldwater-Nichols Defense Reorganization Act into law (PL 99-433). This act, partly a response to lack of coordination at the Joint Chiefs of Staff level during the failed U.S. hostage rescue effort in Iran (Operation Evening Light, 1980), reshaped the command structure of the services, placing emphasis on "joint service" of military officers. A major long-range effect of the reform was to substantially shift authority and control from uniformed officers to civilian appointees. One goal was to eliminate duplication between the services through "Tri-Service Reliance," that is, the coordination of defense programs and plans for both science and technology and for testing and evaluation among the three services. In some cases, work for all three services might even be co-located at a single facility. As it was implemented, Tri-Service Reliance could directly affect the Navy's research and development centers and laboratories.

During the Reagan administration, efforts to reduce the federal bureaucracy by transferring to the private sector R&D work of government-owned facilities resulted in more work contracted out and some increase in the use of academic institutions and commercial firms in defense areas. President Reagan had appointed a White House Task Force, as well as an independent group, the Grace Commission, to identify ways in which federal activities might be better handled by the private sector. In the area of naval research facilities, however, no major government-owned laboratories or experimentation facilities were converted to private ownership or operation.

In addition to privatizing efforts, a parallel emphasis on transfer of science and technology from the defense sector to the commercial sector continued. Both the Stevenson-Wydler Technology Innovation Act of 1980 and the Federal Technology Transfer Act of 1986 sought to expedite such transfers. Reagan signed an executive order in 1987 to encourage cooperation among federal laboratories, industry, and universities. The National Competitiveness Technology Transfer Act of 1989 regularized and expanded the use of cooperative research and development agreements (CRADAs) between DOD and Department of Energy (DOE) laboratories and the private sector. Between 1988 and the end of 1992, the Navy had entered into ninety CRADAs.[36]

The Packard Commission values of reducing bureaucracy, the Goldwater-Nichols principles of centralizing defense, the Grace Commission emphasis on privatization, and the principles of international competitiveness and technology transfer continued to affect R&D policy in various ways, even as the international situation changed. The policies which had been conceived and developed during the last years of the Cold War persisted with their own momentum as the basis for management practices during a period that demanded new thinking and new policies. Many new and responsive policies were suggested at different levels; however, no single philosophy or doctrine emerged as the guiding principle for reshaping defense or Navy RDT&E.

Many factors prevented the emergence of such guidance. Defense administrators seemed at first hesitant to interpret the events in Eastern Europe as showing a decline in Soviet influence or as likely to produce a reduction of classic Cold War international tensions. The arms race with the Soviet Union had driven the U.S. defense budget since the late 1940s, and that established pattern of thought was not easily abandoned. Through 1986 and 1987, many planners in Congress and the Defense Department assumed that reforms taking place in the Soviet Union would have the long-range effect of strengthening the Soviet social and economic system. Some expressed concerns that the United States was beginning to lose its "technological edge" in weapons and in general defense preparedness.

International events in 1989 and 1990 were so dramatic that the diminished Soviet threat could not be denied. Immediate demands from the public and from political leaders to reduce defense expenditures became the most important consideration in RDT&E management. Quite suddenly, planning and management efforts shifted from the priorities established in the Reagan years to issues of base closure and budget cutting.

The abrupt public and political recognition of the decline in Soviet threat left defense planners generally disoriented. For the Navy's community of RDT&E centers, which represented a diverse set of technical facilities and specialties serving a variety of customers, the rapid collapse of old priorities only accentuated the lack of central direction that already existed. The Navy's RDT&E community went into a "dither state," according to one senior technical manager.[37] This meant that the facilities, over the period 1985–1993, faced the demands of multiple transitions without developing any central plan or coordinated, systematic response to the emerging post-Cold War environment.

4

SPAWAR as a Potential RDT&E Coordinator

Many naval officers and naval civilian employees working in the management of naval R&D centers attributed the failure to respond in a concerted fashion to the crises and transitions of the late 1980s and early 1990s to an *organizational gap*, a structural flaw in the system of naval management. The gap was simply that no office in the hierarchy of naval structure had been *tasked* to express the collective viewpoint of the RDT&E centers. A number of naval officers and civilian naval technology experts lamented the fact that an effort to create such an office within the Space and Naval Warfare Systems Command in the late 1980s had not been successful.

As part of the Packard Commission-inspired effort to reduce the size of the naval bureaucracy, Secretary of the Navy John Lehman had closed the Office of the Chief of Naval Material in 1985.[38] Director of Navy Laboratories (DNL), who had a coordinating role for the RDT&E centers and who had reported to the Chief of Naval Material, was at first transferred to the Office of the Chief of Naval Research and then, in 1986, was moved to the new SPAWAR command. That command, created out of the existing Naval Electronic Systems Command, was charged with the added responsibility for developing an overall "warfare systems architecture."[39] This term, in the language of systems analysts, implied the coordination of all systems of strike, surface, air, and subsurface technologies. Such responsibility would include translating changes in mission and warfighting strategy into a coordinated plan for technological development. In short, the architecture mission of SPAWAR held the potential to develop a central response to external developments from the naval R&D centers' point of view.

However, in its formative years, SPAWAR found it difficult to convert from a systems command limited to a specific procurement responsibility in electronics to a centralized R&D planning office as well, and centralized R&D planning eventually atrophied. In 1990–1991, a series of oral history interviews focused on the recent history of SPAWAR and the reasons it had not fulfilled its mission to deal with the Navy RDT&E planning gap.

During these interviews, participants and observers attributed the failure of SPAWAR in its systems architecture role to a variety of factors. Most agreed that SPAWAR never received the necessary manpower and budget and that the drawdown of the Navy bureaucracy caught up with this office before it had a chance to properly organize.[40] Some believed the individual naval officers in charge of SPAWAR did not have the seniority and breadth of vision necessary to bring a comprehensive systems approach to bear on R&D planning and, at the same time, lobby for more resources to accomplish the task.[41]

A number of individuals who had either worked within the office of the DNL and SPAWAR or who were closely familiar with the DNL's functions attributed the lack of a systems planning role to specific bureaucratic factors, such as the weakening and eventual elimination of the central leadership post of the DNL.[42] Retired Rear Admiral Wayne Meyer pointed to the assignment of a recently appointed three-star admiral, Glenwood Clark, to head SPAWAR, a post requiring a higher rank. Meyer also noted the natural competition between the command's responsibilities in electronics and its added systems responsibilities.[43] Representatives of the management consulting firm of Booz, Allen & Hamilton emphasized that the high rate of turnover of naval officers broke any continuity of leadership.[44] Persistent planning responsibilities in the Office of the Chief of Naval Operations (OPNAV) and elsewhere led to competitive voices, regarding central planning, diffusing any effectiveness of SPAWAR in this area.[45] On top of those problems, the understandable tendency of each diverse R&D center to focus on its immediate problems and specific budgetary issues worked against central coordination. Gerald Schiefer, who served as the last DNL from October 1989 to January 1992, particularly lamented that people made decisions on "their own little tiny piece" and that there was no overall Navy Systems Engineer.[46]

Several naval managers suspected that when Secretary Lehman established SPAWAR in 1985 and charged it with the warfare systems architecture and engineering responsibility, he may have been simply responding to critics who believed a systems approach would solve planning problems, without himself endorsing or recognizing the merit of such a plan. Several observers, most notably retired Admiral Meyer, spoke forcefully about the lack of support for the systems architecture mission from the Office of the Secretary of Defense. Meyer was indignant about several issues: poor management design with too broad a span of control, assignment of inexperienced or junior-ranking officers to responsible positions, the political nature of the Joint Chiefs under the Goldwater-Nichols Act, and what he saw as widespread failure to understand the nature of systems architecture.[47]

The Director of Navy Laboratories had served as a coordinating and advocacy position for the laboratories since the establishment of the post in 1966. That office had been somewhat downgraded after its inclusion in the

SPAWAR systems command, and its coordinating and leadership role had been weakened there. Then, in 1992, the Secretary of Defense eliminated the DNL position in implementing the warfare center concept. With the transfer of responsibilities for planning and coordination from the DNL to the various offices in the Navy systems commands, the RDT&E community within the Navy no longer had a single individual representing its interests at a high level at the very time that consolidation and closure went forward.

The internal or organizational explanations of the lack of response to multiple transitions faced by the Navy's RDT&E community through this period show why particular offices or bodies did not develop as a locus for central planning and advocacy. All of the explanations offered by the interviewed personnel as to the role of SPAWAR appeared to hold merit. By the early 1990s, however, it was clear that SPAWAR, though it continued to function as a systems command in the C^3I area, had not developed a major role in responding to the need for overall coordination and planning. The issue of exactly why that mission did not develop became moot.

Multiple external transitions presented the R&D centers and laboratories of the Navy with a series of conflicting forces which, more than ever, required some sort of concerted and organized scheme of response. It was a period of confused seas, and the R&D centers faced those seas without the benefit of a single designated advocate, a single pair of hands on the wheel.

From a broader perspective, the naval RDT&E community was not alone with its problems. Despite the collapse of the Soviet threat, most United States naval authorities continued for a few years to base policy and doctrine on that threat. By 1993 planners on the Fleet side of the Navy, in OPNAV, had only started to define new warfighting priorities and to attempt to convert them into doctrinal needs. Throughout 1992 and 1993, various offices began to think through the future shape of the Navy, future warfighting requirements, and new ways of structuring technology procurement. Within OPNAV several groups in the Office of the Deputy Chief of Naval Operations for Resources, Warfare Requirements and Assessments (N 8) focused on future needs.[48] Nevertheless, the offices on the operational side of the Navy did not have an administrative structure or a central channel through which to translate new thinking about warfare styles into planning for R&D. Furthermore, the CNO did not have the power to set the Navy's mission; since Goldwater-Nichols, this power resided at the level of the Secretary of Defense, with advice from the Joint Chiefs.[49]

At an even higher level, leadership and direction regarding the post-Cold War international environment were not at all consistent or clear through the early 1990s. The lack of direction within the Navy was compounded by the fact that the President was responsible for setting a broad defense policy direction, based upon advice from his appointed Secretary of Defense. As a result of the 1992 elections, plans regarding defense force levels abruptly changed from those laid down by President George Bush on advice from Secretary of Defense Dick Cheney to the plans put forward by President Bill Clinton and his first Secretary of Defense, Les Aspin.

Furthermore, in the first year of the Clinton administration, there appeared to be unresolved differences between the announced position of the Secretary of Defense and the White House.

As Navy managers sought a clear policy, they encountered a vigorous and lively policy debate at the command level, with many voices often in disagreement. Policy and planning documents issued by significant agencies, commissions, executive offices, and panels during the period outlined the varied responses and the directions. The documents discussed in the following pages are a fraction of the total volume of reports and papers dealing with science and technology policy issues, but the ones chosen represent the spectrum of opinion and the gradual recognition of the changing international environment and how it might affect the naval RDT&E centers. These government and military responses are described against the background of the unfolding dramatic changes in international relations.

5

1986–1989 Reorganization and Reform

As mentioned earlier, the Goldwater-Nichols Defense Reorganization Act (PL 99-433), passed by Congress in 1986, amended U.S. Code, Title 10, the basic law spelling out the organization of the armed forces. The six titles of the act required a massive reorganization of several features of the Defense Department. The act particularly spelled out the duties and functions of the Joint Chiefs of Staff, established a requirement for joint-duty service by military officers, required a cut of some forty thousand civilian and military positions, and required the JCS Chairman to prepare a study of the roles and missions of the various military services every three years.

In 1986 Secretary of the Navy John Lehman retained the consulting firm of Coopers & Lybrand to study the U.S. Navy shore structure, including its laboratories, as part of his plan to streamline and modernize the Navy's organization. This study, "Management Analysis of the Navy Industrial Fund Program: Naval Laboratories Review Report" (June 1986), in the tradition of earlier management studies of naval shore facilities, focused on six functional areas: Operations, Organization Studies, Financial Management, Management Information Systems, Procurement and its effectiveness and efficiency, and Personnel. The team visited sixteen naval centers and laboratories and conducted over one thousand interviews to examine the Naval Industrial Fund (NIF) research and development facilities. Although the team found the individual facilities well managed, it decried the lack of common or cohesive strategy among the laboratories and centers. In general, Coopers & Lybrand criticized the high proportion of budget devoted to management and contracting services in labs as opposed to direct work, echoing the Packard Report and earlier studies that had criticized the growth of the management side of R&D.[50]

The Coopers & Lybrand report supplemented the existing ideas of the Packard Commission and provided a basis for eliminating the Chief of Naval Material and transferring R&D coordination as well as the Office of the DNL to the Office of the Chief of Naval Research and, later, to

Secretary of the Navy John Lehman carried out many of the cost-cutting recommendations of the Packard Commission Report during the Reagan administration.

DN-SN-86-10019

SPAWAR. There was no perception in the Coopers & Lybrand report that recent events in the Soviet Union would produce an altered strategic environment requiring any rethinking of R&D priorities. Although the report was dated and released in June 1986, the research and writing had been conducted during 1984–1985. Mikhail Gorbachev was elected Secretary General of the Communist Party of the USSR in March 1985, and the Reykjavik Summit, which gave a foretaste of the forthcoming disarmament agreements, was not held until October 1986, several months after the report was finalized and issued.[51] Thus, the Coopers & Lybrand report represents a continuation of earlier efforts, like those of the Packard Commission, to examine RDT&E facilities with an eye to strengthen them by reducing excess overhead, assuming the Cold War would continue to provide the fundamental strategic situation.

The Coopers & Lybrand findings and recommendations specifically focused on strengthening the facilities. They found "no clearly articulated corporate strategy directing the NIF RDT&E community" and noted that there was no "established, regular, formal process for evaluating the performance" of the facilities. Coopers & Lybrand determined that constraints on personnel management worked against "effective and timely" completion of both technical and administrative work. Their report also concluded that procurement authority and guidance were uncoordinated and contradictory. Furthermore, the "stabilized rates" used by NIF facilities in internal charges against program budgets were inappropriate for the unpredictable nature of R&D work: they were set by fiat two to three years in advance of the fiscal year in which they were to apply. Final costs reflected changes which, in most cases, were required by further administrative rulings. Yet NIF facilities were powerless to adjust to the changed reality. The report listed specific improvements needed in the R&D structure particularly in the management information systems that varied from center to center and required coordination to become effective. In straightforward language, their report recommended improvements to address all of these organizational, procurement, acquisition, and information issues.[52]

In the summer of 1987, the Defense Science Board (DSB) conducted a study of the technology base management, which was published in December of that year. In "Technology Base Management" the board identified several issues, expressing concern with top management's lack of attention to the Technology Base and with the decline in quality, and remarked that the system seemed deficient in making the transition from research into products and services in operations and deployed weapons.[53]

Like the Packard Commission and the consulting firm of Coopers & Lybrand, the 1987 Defense Science Board assumed that the Cold War provided the strategic basis for future planning. Echoing decades of such an approach, the DSB report viewed the situation with alarm: "Soviet weapons system performance approaches and in some cases exceeds that of U.S. and Allied forces." To correct these deficiencies, the study recommended several reforms that included establishment of an Acquisition Executive in each service and revitalization of 6.3a technology demonstration projects.[54]

The DSB was concerned that the United States was losing its technological advantage; that it was getting less than full value for its R&D dollar; that inefficient overlap between commercial and defense research existed; and that reorganization underway because of the Packard Commission Report and Goldwater-Nichols legislation had left some turmoil in the services. The board noted that its recommendations would encounter institutional and political opposition and resistance, but that the best way to meet opposition was to continue with experimental programs and demonstration projects which showed the validity of new approaches.[55]

The Defense Science Board also posed the problem of poor leadership in U.S. defense R&D against the high quality of Soviet arms. At this date the Soviet threat still provided the basic strategic orientation of such planners. Gorbachev's reforms of "*glas'nost*," or openness, and "*perestroika*," or restructuring, were viewed not as harbingers of instability, but as reforms that could strengthen the Soviet political, military, and strategic capability to threaten the West. The board decried what it saw as poor performance by U.S. government employees, the practice of competition with the private sector, and ineffective leadership at the top, all of which led to a lack of quality in weapons technology.

The report of the Defense Science Board reflected themes characteristic of the Reagan administration: the virtue of the private sector over government-managed and government-regulated facilities and the Soviet threat that required strengthened defense and improved defense procurement. The board echoed the Packard Commission's plea for more efficiency and less bureaucracy. In December 1987, when the summer study was published, Soviet forces had not yet begun their withdrawal from Afghanistan, and the promise of disarmament, suggested at Reykjavik in the fall, seemed unlikely.

The DSB study in the summer of 1988 reflected a shift to a concern with economic competitiveness. "Defense Industrial and Technology Base," written by a group chaired by Robert Fuhrman, stressed the need to keep defense planning in touch with strategic and international economic competitiveness:

> the defense industrial and technology base faces new and difficult challenges in the current and expected world market. We found that the defense business is now truly global. America and its allies are interdependent in many industrial resources essential to national security. Furthermore, America faces an increasing loss of technological leadership to both our allies and adversaries.

In contrast to the Coopers & Lybrand (1986) and the DSB (summer 1987) studies, the 1988 summer study focused on international economic competition and the global nature of technology rather than simply reiterating the Soviet threat that required a strong response. To an extent, the United States' lack of international competitiveness appeared to be a more serious threat than communist military aggression. And by 1988, DSB planners knew that international priorities were changing, even though deterrence would remain the basis for strategy.[56]

The 1988 DSB report accepted the traditional Cold War rationale for excellence in weapons technology: "Our national security is based upon a strategy of deterrence. We have chosen not to match our adversaries soldier for soldier and bullet for bullet. Instead we have chosen to maintain a degree of technological superiority sufficient to overcome our numerical disadvantage." But problems arose: "America's technological superiority has diminished. Many countries, including Japan and the Soviet Union, challenge our leadership in technologies essential to defense." And there were other threats: "Europe is on the verge of a planned economic unification. The Pacific Rim nations are pressing economic expansion . . . in the global electronic and defense markets."[57]

The changes were profound, and "the days of Fortress America are past. We are, and will remain dependent on foreign resources for critical components of our weapons systems." Although foreign dependence could not be eliminated, a strong federal effort could offset the "loss of leadership in key defense technologies."[58]

Fuhrman made several recommendations, particularly focused on problems of lack of competitiveness and modern shipbuilding capacity, questions of ethics in contracting firms, and the over-use of the best-and-final-offer system. The study suggested a need for further federal funding of the private sector, which could not engage in capital investment for defense when such an investment often interfered with a company's cash flow.[59] In some regards, the 1988 summer study's emphasis on government procurement appeared to offset the 1987 emphasis on the private-sector. Fuhrman suggested that private enterprise was not capable of resolving defense R&D deficiencies by itself, and that government leadership was required for the United States to remain competitive in the international environment of technological advance. The 1988 study represented a shift in basic ways from "Cold War" thinking, without completely abandoning the Cold War rhetoric and the concern with Soviet power.

Two months after the 1988 summer study was published, Gorbachev announced a planned withdrawal of 500,000 Soviet troops from Warsaw Pact nations in Eastern Europe. This decision opened the way to the breakdown of Communist Party control of the satellite states over the next year. Even in the face of this clear change, many experts in both academia and government continued to disagree as to whether the changes were significant; some assessed the Soviet developments as fundamental, while others assumed they were designed to "lull" the West into a false sense of security.[60]

At the time of these international developments, the Office of Technology Assessment prepared a study of the technology base, or Science and Technology (S&T), in 1988. Reflecting the concerns in the title, "Holding the Edge: Maintaining the Defense Technology Base," the congressionally sponsored report concentrated on defense S&T programs and laboratories and on ways to ensure the introduction of technology into specific systems.[61]

OTA decried the Defense Department's lack of a "centralized system for strategic planning" for technology base programs. DOD had a federated rather than a centralized system, with the Under Secretary for Defense Development and Research playing a coordinating, not a decision-making, role.[62] OTA viewed this approach as ineffective in developing a coherent program in the fields of science and technology. While the federated system allowed each service to plan its own work, the lack of centralized planning and direction seemed inefficient to the OTA staff. Indeed, OTA identified the absence, at the DOD level, of the sort of systems architecture role anticipated, but not developed, in the Navy at SPAWAR.[63]

The Office of Technology Assessment generally opposed the idea of in-house, government-owned and -operated facilities. It spelled out management deficiencies: problems of recruitment and retention of qualified personnel, inefficient conduct of daily business, and a poor record in updating equipment and buildings that rapidly became obsolete with rapidly advancing technology.[64] OTA recommended examining several other means of organizing government research besides government ownership, including government-owned, contractor-operated (GOCO) and totally private facilities. The office argued that private contractors and private firms had more flexible personnel rules and could move faster and more efficiently than government agencies in purchasing equipment. OTA did admit that a virtue of in-house facilities was their ability to move quickly, under direct orders, to adapt to operational-side demands, while private sector operations could respond only after cumbersome procurement steps had been taken.[65]

To an extent, the OTA report recognized that international changes were bringing about a shift in defense industry priorities in other countries. The Office of Technology Assessment found evidence in other developed countries of a reduction in research directed to defense alone and an increase in research that could apply to both civilian and military functions. Furthermore, OTA stated that there was a general international movement away from government-owned laboratories towards a market-driven civilian sector. OTA urged greater DOD access to commercially developed technology and a greater awareness of the international nature of the private sector.[66]

Using the private sector entailed several difficulties. Companies engaged in both civilian and defense work were not always interested in pursuing the military side; many high-technology firms were moving abroad to better meet foreign competition; and with the complexity of modern technology, the process of getting from raw material to finished product repeatedly crossed international boundaries. The Defense Department would have to adjust to this new world. In 1988, OTA, like the Defense Science Board, focused on international developments, recognizing international competitiveness and the changed world as directly affecting the United States' ability to defend itself. Neither group was overly optimistic about the changes in the Soviet Union.[67]

Reductions in the U.S. defense budget had been discussed publicly after Gorbachev's announced cuts in troop strength in 1988. In November 1988, Prime Minister Margaret Thatcher stated in a Washington, D.C., speech that the Cold War was over, and a year later, Secretary of Defense Cheney appeared to concur. By November 1989, a *Washington Post* reporter editorialized that the "disintegration of the Soviet-led Warsaw Pact alliance is outrunning the conventional arms-control process." By 27 November the phrase "peace dividend" had entered common parlance to describe potential defense budget cuts.[68] Within DOD there was a clear understanding through mid-1989 that the congressional pressures for budget cuts would need to be addressed. From mid-1989 through mid-1990, U.S. journalists and defense planners began to assume that the Soviet threat was diminishing. As Rear Admiral Wayne Meyer noted, the offices of the Secretary of Defense and the Joint Chiefs were highly responsive to political pressure, and political pressure now mounted to cut the defense budget.[69]

6

1989: The Cold War Ends

A cascade of events in 1989 and 1990 suggested that a profound change in the world order was underway. In particular, in early 1989 the Soviet Union held its first multi-party elections, and non-Communist parties did surprisingly well. In mid-1989 the announced troop removals from Eastern Europe began. That fall, Hungary opened its borders, and East Germans took the opportunity to emigrate to West Germany through the open Hungarian borders. In December 1989 the Berlin Wall was torn down. Democratic elections held in Poland were won by non-Communist parties; similar elections in East Germany, Hungary, Romania, Bulgaria, and Czechoslovakia followed in 1990. In October the two Germanies agreed to reunification. These events provided the international setting as reform and restructuring of the U.S. Department of Defense proceeded through 1989 and 1990.

In the United States, from 1986 through mid-1989, the Department of Defense had already undergone considerable reorganization, following some of the recommendations of the Packard Commission and implementing the requirements of the Goldwater-Nichols Reorganization Act. In February 1989 President Bush directed the Secretary of Defense to "develop a plan to accomplish full implementation of the recommendations of the Packard Commission" and to review Defense Department management overall. In "Defense Management: A Report to the President" (July 1989), Secretary Dick Cheney reviewed some of the changes, indicating that there was "no basis for complacency" but examining in detail "progress to date."[70]

Quoting guidelines from the Packard Report, Cheney spelled out what had been done and what needed to be done to make the services' logistics and purchasing more efficient and less wasteful. In particular, Cheney detailed the large scale of the Defense Department's acquisition work force, some 580,000 civilian and military personnel. The report aimed at improving the training and quality of that work force and sought a reduction in its size.[71] Cheney also noted the vast amounts of time and money spent in responding to congressional requests for information. With what may have been a sign of frustration, or at least an effort at irony, he recommended that Congress itself charter a study of ways in which the relationship be-

Secretary of Defense Dick Cheney, under political pressure to reduce defense costs with the end of the Cold War, directed implementation of management reforms throughout the Defense Department.

NHC L-file

tween Congress and the Defense Department could be made less time-consuming and wasteful.[72]

As part of his management reform effort, Secretary Cheney requested the service secretaries provide by 1 October 1989 specific plans reflecting how the suggested guidance on cutbacks in his report to the President would be implemented. The Defense Management Report recommended that the scale of defense facilities be cut back through cooperative programming, through extensive cooperative use of facilities, through closure of facilities, and through consolidation of activities within and among the services. This emphasis on inter-service coordination and cooperation came to be called "Tri-Service Reliance."

As some individual facilities and offices merged, the resulting inter-service units were colloquially referred to as "purple," reflecting the presumed color blend of the three services' uniforms.

Despite the unfolding events in Europe, most analysts continued to rely on the pattern of seeking improvements in the defense research and development establishment based on a concerted Soviet threat. For four decades, the Cold War had provided the basis for planning, and it was difficult for defense policymakers to immediately develop a new set of premises upon which to base projections. The October 1989 DOD study viewed with alarm the continued threat from the Soviet sector, documenting its concerns from several other recently published studies such as the 1987 Defense Science Board summer study and the 1989 Office of Technology

Assessment report, "Holding the Edge." In the executive summary of the OTA report, the authors had called for greater integration of the three services within DOD, better concentration on taking the products of research into the field and the fleet, and a planned long-term investment for science and technology.

Like the OTA study, the DOD report expressed concern over the productivity of the defense laboratories. To establish leadership to deal with R&D issues, the report recommended an executive position in the Office of the Secretary of Defense to report to the Under Secretary of Defense for Acquisitions to give guidance and oversight.

The 2 October 1989 report "DOD Management of Technology Development—Implementation of Recommendations of Defense Management Review" noted that the current situation demanded "flexibility to respond to changing military threats," thereby implying some recognition of recent international developments. However, the implementation statement included among those threats the possibility of a "revitalized Soviet." In addition, like the 1988 DSB report, the October 1989 DOD report noted the problem of international "economic competitiveness" as well as the Soviet threat.[73]

The DOD report identified the "faltering US technological position on the world stage" and repeated the phrase, suggesting that the faltering position, both in the military sense and in economic competition, "compels the most vigorous action which is prudent." The vigorous action suggested in the report was the establishment of a central S&T authority in the Defense Department.[74]

The report regarded the government-owned, contractor-operated laboratories of the Department of Energy as holding promise for a better model than in-house laboratories. It alluded to possible future Soviet strength or other changed military threats without further definition and saw U.S. weapons technology as not prepared to adjust to the changes underway. GOCOs, as semiprivate entities, would have more flexibility for adjustment. The October 1989 report on implementation of management reform did not reflect a close familiarity with the range of problems encountered by former CNO Admiral James Watkins as he sought to reform and rein in the GOCO national laboratories during his tenure as Secretary of Energy from 1989 to 1992. Admiral Watkins, on his appointment as head of the DOE, announced that he intended to establish a "new culture" there. But he found that the old patterns of independence and procurement and ways of handling environmental issues persisted stubbornly at the DOE national laboratories and GOCO industrial sites.[75]

Meanwhile, Secretary Cheney's request for review and implementation of the Packard Commission and Goldwater-Nichols changes provoked other responses. The Department of the Navy Management Review Task Force issued "Plans for Initial Implementation of the Defense Management Report" on 1 October 1989, and it was approved for implementation on 11 January 1990.

As a result of the Defense Management Report, the DOD comptroller issued several decisions aimed at achieving savings in the 1990 and later year budgets. For example, Defense Management Report Decision 922 proposed a significant savings in budget for the years 1991–1995 by consolidating RDT&E activities to reduce overhead, streamline operations, and centralize professional staff associated with specific technology areas. In a memorandum entitled "Consolidation of R&D Laboratories and Test Facilities," Deputy Secretary of Defense Donald J. Atwood held off implementing the comptroller-recommended actions but set up teams in November 1989 to review the recommendations and to develop detailed implementation plans. The teams were to complete their work and report back by 1 May 1990. These studies chartered by the Secretary of Defense were to focus on inter-service consolidation rather than on the intra-service consolidation in which the three services were already engaged. In this specific administrative procedure, Secretary Cheney instructed the services to become more efficient in acquisition through detailed implementation plans for tri-service coordination of programs, planning, and some co-location of facilities.

One inter-service working group focused on both research and development activities and test and evaluating activities. In the case of the Navy, many RDT&E activities were "full-spectrum," that is, operating over the whole range of research, development, test, and evaluation of a particular technology coordinated through a single program and often managed through a single center. In Navy parlance, full-spectrum also meant that facilities often assisted both the manufacturer and the naval customer in working out details of production, providing service to the fleet as systems were installed and debugged, and assisting in technical support of the systems through their service lives. In effect, full-spectrum referred to support of Navy systems from concept through retirement, or from "cradle to grave."

The working group requested detailed information from all potentially affected service headquarters and field activities through December 1989 and January 1990.[76] The Office of the Secretary of Defense spelled out ten specific criteria on which to judge whether a candidate facility should be closed or consolidated with those of another service:

1. Appropriateness, value, and affordability of similar activities
2. Expectation of similar effort in commercial sector
3. Level of effort and quality that had to be maintained
4. Relevance to mission
5. Critical mass of personnel
6. Appropriate balance of in-house vs. contracted-out work
7. Basis of current location related to customers, test facilities, and academic facilities
8. Impact of present and future geographic and environmental considerations

9. Impact of future changes in missions and technology thrusts
10. Investments and people-related costs

Using these ten criteria for deliberating on consolidation or closure, the Secretary of Defense groups were to consider possible actions ranging from complete closure of an activity through reduction, privatization, partial elimination, and conversion to a government-owned, contractor-operated facility, and selective establishment of lead laboratories.[77]

In response to the structured request by Cheney, the Secretary of the Navy ordered the newly appointed Director of Navy Laboratories, Gerald Schiefer, former technical director of the Naval Weapons Center, China Lake (1986–1989), and the Chief of Naval Research, Rear Admiral J. R. Wilson, to head a panel examining the Navy's facilities with an eye to possible consolidations and economies.[78] At first, little room for consolidation was found; few economies could be realized in the "near term." However, some areas were developed for future study and some possible management economies were identified. In particular, aligning several field activities with similar or related responsibilities in a single field organization and placing such organizations under a single headquarters organization suggested a way to improve service and accountability. A pattern of organizing activities along warfare-function oriented lines emerged from the study, creating "megacenters" that clustered related smaller facilities around existing RDT&E centers.[79]

In studying this approach, the working group recognized that its conclusions reflected some that had been produced years before, particularly noting the December 1977 Gavazzi Study, officially published as "Office of Chief of Naval Material Draft Report of the CNM Ad Hoc Group on Functional Realignment."[80] The Gavazzi Study reflected efforts in the Navy since the mid-1960s to consolidate and cluster laboratories to achieve what was regarded as an appropriate "critical mass," or large scale, appropriate to R&D activity. Schiefer later recalled that his group came to recognize that long-standing ideas about consolidation, often recommended in past reports but never achieved, could now be implemented under the pressure of the "drawdown."[81]

The initial 1990 Navy study, or "Phase I" work, was completed in February of that year; Richard Rumpf, Acting Assistant Secretary of the Navy for Research Engineering and Systems, directed that this study be expanded to consider all naval shore activities engaged in research, development, acquisition, and fleet support. To produce this combined response by 1 March 1990, Rumpf appointed a "tiger team" headed by Gerald Schiefer; the team's activity was designated Phase II of the Navy's response to the request from the Office of the Secretary of Defense.

This Phase II response took a broad view, linking management decisions to international changes and looking closely at a set of "knowns" and "unknowns" in predicting what should be done about consolidations and closures of RDT&E and fleet-support facilities. Unfortunately, the tiger team saw no clear consensus on what the future force structure of the Navy should be. The team would have preferred to start with a quantified force structure and then work "top-down" to determine what facilities would be required to go forward to support that structure. Instead, it had to make certain assumptions about what minimum sets of activities would be needed by the Navy to carry out its future roles in research, development, and acquisition. In this manner, the tiger team addressed fundamental questions about the nature of the present and future U.S. strategic and military situations without the benefit of clear input from higher authority. The team almost had to make guesses about national policy in order to find a starting point. In effect, its report reflected the kind of centralized planning role that had been written into the SPAWAR charter several years before but had not been effectively executed.

The team plunged into geopolitical and fiscal issues and produced an analysis based on those elements they could agree were "knowns" in the welter of "unknowns." As given factors, they noted that events in Eastern Europe had changed the mission of the Navy and that the consequent demand for a "peace dividend" would reduce defense dollars. The Office of the Secretary of Defense would restructure the Navy if the Navy did not do it for itself. There would be more emphasis on "low intensity conflict." Although low intensity conflicts might be less threatening to U.S. security than a potential nuclear exchange with the Soviet Union, such conflicts were highly unpredictable and could come from a variety of sources.[82]

In describing the warfare environments that the Navy might encounter, the team considered these alternatives: continued reliance on strategic deterrence, drug interdiction for the military, and increased low intensity conflict. Low intensity conflict would require "uncertain alliances for the U.S.," precision action, and a reduced public tolerance for battle casualties. Furthermore, the future held out probable withdrawal from overseas bases, or "forward sites," fewer weapons platforms and weapons systems, and an uncertain threat. Throughout its discussions, the team assumed "all existing Navy missions will remain relevant," with possible changes of emphasis within the missions.[83]

The team also observed that the roles of the systems commands would be diminished under the Defense Management Review, while the roles of program executive officers would expand. High reliance on the technological superiority of U.S. warfighting systems would continue.

In a structured manner, the tiger team reviewed potential warfare environments, the likely developments of technology—the "technical environment," the changing acquisition environment, and the nature of the "support environment," which would be more multi-service oriented and smaller in scale. Using these considerations, the team developed a picture of the near future and a sense of the generic functions that would have to be carried on in research, development, acquisition choices, and support for technology in the Navy.

These background discussions served as a basis for considering consolidations and closures. The team moved to a careful analysis of ways to reduce or eliminate duplications by "aggregating activities" into a smaller number of centers, giving them all full-spectrum responsibility. Using the principles developed, the team generated a list of sixteen questions comprising a "consolidation reality test." This analysis would determine if a proposed action saved money, reduced duplication, favored environmental protection, met the criteria of providing full-spectrum service and of facilitating technology transfer, and meshed with other aspects of the administrative and military environment envisioned by the team. The team then used this test to consider proposed consolidations, generating a list of closures, mergers, and inter-service transitions. The report produced specific and objective recommendations, reflecting sophisticated methods of analyzing and structuring the decision process.[84]

This analysis in early 1990 reflected more thinking about the long-range strategic impact of the changed geopolitical situation than had the earlier reports issued between 1986 and 1989. Nevertheless, the 1990 Phase II tiger team report could not assume any specific research or development implications of the changing requirements for weapons systems since no such operational requirements had yet been determined. Nor did the report receive wide or public distribution; it was simply forwarded through advisory channels as part of the planning for base closure.

7

1990: Naval RDT&E Recognition of Changed Priorities

In April 1990, Gerald Cann, Assistant Secretary of the Navy for Research, Development, and Acquisition, forwarded to the Under Secretary for Defense (Acquisition) an extensive memorandum entitled "DOD Test and Evaluation, Research and Development Facilities Paper" (12 April 1990). This memorandum outlined the Navy's evaluation of "study options" on how it should restructure its RDT&E facilities in light of the changes in domestic and international priorities.

The report was a formal response to the Defense Department's call for consolidation, which had been summarized in DMRD-922, issued in December 1989. In effect, the Navy informed DOD of its collective preferences on reorganization of RDT&E using the conclusions and careful recommendations of the Phase II tiger team study. The Navy accepted the concept of some tri-service consolidation, as long as the changes guaranteed retention of specifically "service-unique" maritime programs in the Navy. The Navy recommended establishing a DOD-wide approach similar to the Navy's system of industrial funding, but opposed establishing a new layer of bureaucracy in the Office of the Secretary of Defense, as that move would counter the efforts to streamline as suggested in DMRD-922. In general, the efforts to reduce duplication received a warm endorsement from Cann.

To an extent, Assistant Secretary Cann and DNL Schiefer reflected an effort by the material side of the Navy to respond to the changing world environment. Cann's memorandum showed thoughtful analysis of the changed international situation and summarized the thinking developed in Schiefer's Phase II study. For forty years the Navy had "placed its highest premium on deterring and defending against possible Soviet military aggression." Even with that situation changed, "technological superiority" would continue to be "an essential element of the future national strength and necessary for future naval superiority." The report noted that "missions assigned to naval forces are becoming relatively more important in the overall national security posture, notwithstanding the diminishing threat to Europe from the USSR." The nuclear deterrent side of that role would continue to be important in a "post-START world." Considering situations such as Libya and Panama, the April 1990 report stated:

> The presence and crisis-response missions, as well as naval contributions to other national security concerns, such as drug interdiction, seem no less likely in the future than in the past to require forward-deployed and expeditionary naval forces capable of achieving military success against the more sophisticated weaponry now coming into third world inventories.[85]

Although articulating that response to the changed environment, Cann proceeded to outline how restructuring and realigning the Navy RDT&E community would lead to four full-spectrum RDT&E centers: Naval Air Warfare, Naval Surface Warfare, Naval Undersea Warfare, and Naval C^3I, and a fifth facility, the "Corporate Laboratory," or Naval Research Laboratory. Cann's report lacked details on the thinking or the methodology of reality-testing first outlined in the Phase II work, nor did it show how the strategic implications of limited intensity conflicts would specifically influence the closures and consolidations envisioned. In Cann's 1990 report, all naval RDT&E activities would divide into streams of "like-purpose" activities; workloads would drive consolidations and closures. A combination of the industrial funding premises and central direction would encourage the most efficient use of facilities within the budget available. The report spelled out the main functions of the streams, several potential consolidations, potential tri-service activities, and some potential long-term consolidations.

To guard against draconian cuts in facilities, the report showed that basic functions of Navy RDT&E facilities were best achieved by continued operation on an in-house basis. The arguments given were: the need to be a "smart buyer," the need to maintain a technological capability in areas of limited interest to the commercial sector, the ability to provide a quick response in time of crisis, the ability to infuse the "art of the possible" into defense planning, and the need for cradle-to-grave capability to carry technology from development through application and fleet support. Each line of argument was explained and illustrated with extensive specific cases.

The report spelled out the staff size at thirty-four in-house RDT&E facilities and five university contract research laboratories. More than fifty-five thousand staff were at the thirty-nine facilities represented; immediate plans to reduce staff through attrition and reductions-in-force (RIFs) were on the order of two thousand people.

While these studies of RDT&E cuts went forward, the Joint Chiefs of Staff developed some strategic planning, attracting attention in the popular press. In September 1989, in accordance with the requirements in the Goldwater-Nichols Act, JCS Chairman Admiral William J. Crowe Jr. forwarded to Secretary of Defense Dick Cheney recommendations for changes in the assignment of service functions, analyzing the roles and functions of

Admiral William J. Crowe Jr., Chairman of the Joint Chiefs of Staff, 1 October 1985–30 September 1989.

each military service. Admiral Crowe called for further study of the functions of the armed forces, suggesting that they "need to be revised and updated to more clearly reflect current national military strategy, our efforts to harness technology, and our responses to evolving threats to national security." Although Crowe suggested that service *roles* did not need reexamination but only the allocation of *functions* among the services, several documents issued over the next two years reflected examinations of both roles and functions.[86]

While Defense Secretary Cheney's initiatives in defense management continued to provide a stimulus and a search for means of increased *internal* efficiency along the lines recommended earlier by the Packard Commission and by other management studies, other analysts began to stress the changes which the new *external* world situation would require. The review by Admiral Crowe recognized that geopolitical changes were afoot, and a number of other reports, beginning in 1990, also assessed those changes. The Joint Chiefs of Staff issued Draft Publication 3-07, "Doctrine for Joint Operations in Low Intensity Conflict," in January 1990. The Center for Naval Analyses, a federally funded center devoted to operations research and strategic planning for the Navy, issued a series of reports that focused on the need to develop doctrine, training, and procurement plans around low intensity warfighting scenarios. Robert Berg of the center suggested in 1990 that "evolving threats pose problems that result in new operational requirements." Adam Siegel recommended new courses emphasizing limited

DN-ST-88-01410

Re-flagged tanker *Gas King* is escorted through Persian Gulf waters by two U.S. Navy ships, guided missile cruiser *William H. Standley* (CG 32) and amphibious assault ship *Guadalcanal* (LPH 7), during Operation Earnest Will.

intensity conflict for the Naval War College in *A Brave New Curriculum for a Brave New World*.[87]

Historians often speak of "watershed" events in history, events which define or mark the transition from one age to another. In a broad sense, the whole period from 1985 through 1993 represented such a time of transition. However, a close examinations of DOD and Navy statements and announcements throughout the period reveals that the events of 1990 in particular clarified the perception that the "Cold War" was over and that the nation was moving to a "New World Order." Thus, if a single year were to be designated a watershed point, that year would be 1990.

It was ironic that the day Iraq invaded Kuwait, 2 August 1990, was also the day President Bush announced a post-Cold War doctrine stressing U.S. support for a new world order. Although intended as a major foreign policy initiative, his speech was eclipsed in news coverage by the Iraqi action. Together, the invasion and the President's speech also marked the transition between the passing era of the Cold War and the dawning period of new challenges and responsibilities. When President Bush and Premier Gorbachev announced at the Conference on European Security held in Malta in November 1990 that the Cold War was over, they recognized that the eleven months following the "fall of the Wall" had represented the end of an era. As President Bush aligned support for action against Iraq, a concurring statement that the Cold War had ended from Gorbachev was particularly welcome. Over the next months of Operation Desert Shield, DOD concentrated on consolidating the defense of Saudi Arabia, while the President and the State Department worked on building a consensus for a stronger response to the Iraqi invasion.

DSB Advice in the Early 1990s

Several specialized reports prepared by study groups and consultants for the DOD focused on adapting military procurement to the post-Cold War environment.

A 1990 summer study by the Defense Science Board entitled "Research and Development Strategy for the 1990s" spelled out ways in which promising technologies could be pursued through a development phase at a somewhat lower cost than a full procurement would entail. Instead of going to full-scale manufacturing, a variety of means could be used to prepare the technology for when production was needed. Thus, progress could continue but funds would not be wasted on producing weapons and systems, deploying them to the field or fleet, and then retiring them before they were ever used. Among the procedures recommended were strengthening dual-use areas with the commercial sector, careful review of export control lists, and strengthening technology transfer to the private sector. The DSB task force focused on a five-point investment strategy for technology: develop technology with critical impact and core technologies with pervasive impact; avoid technological surprises; provide for manufacturing and process technology to put in both advanced and affordable systems; promote leveraging of dual-use and multinational technology; and improve the infrastructure.[88]

The summary report of the 1990 DSB summer study was even more focused on means of adjusting technology to the new situation. Recognizing that DOD would face smaller budgets, a smaller defense industry, smaller procurement quantities, fewer new technology starts, an aging inventory of weapons and systems, and continuing rapid technology growth, the DSB recommended six strategies for DOD: concentrate on "Breakthrough Technology" that has the potential for revolutionizing warfare; streamline mainline procurement and use it cautiously; use a "Fastrack Initiative" to rapidly produce fieldable prototypes rather than full-scale runs of new systems; allow "Technology Insertion" to insert key advances into existing systems; foster "Commercial Harmonization" with the private sector and international producers; and finally, in the reduced spending environment, concentrate on "Low Quantity Production" to ensure constant modernization without the expense of full-scale production runs of new technologies. These and similar ideas of flexible production to adjust to the new global environment continued to show up in a variety of advisory reports over the next few years.[89]

1991–1992 Post-Desert Storm Base Closures

After the Gulf War in January and February 1991, the Director of Navy Laboratories attempted to evaluate the performance of the Navy's RDT&E activities in relation to that war effort. In a report entitled "Operation Desert Shield and Desert Storm" (31 August 1991), the DNL stressed some concepts that had applied during the Cold War. The report noted that "U.S. defense policy relies on technological superiority of its warfighting capability over that of its adversaries." The United States, however, would make no attempt to match military enemies man for man or machine for machine. Rather its forces would use the "force-multiplying" effect of technological superiority to overcome the other side's advantage in men and machines. Suggesting that the Gulf War might give hints of new warfighting scenarios, the report noted that "future operations are also likely to involve operations in urbanized areas," in which the use of technology to pinpoint targets and to reduce collateral damage would be essential.[90] The report provided solid arguments for maintaining in-house capabilities by noting that, in peacetime, academia shifts to long-range research goals while the commercial sector shifts to profit-making activities. These arguments emphasized continued excellence in technology supported by an in-house capacity for R&D.[91]

There was no detailed analysis of strategic implications of the new world situation, but there was some attempt to justify R&D, particularly as it related to the performance of weapon systems during Desert Storm. As this document demonstrated, the DNL still sought to develop an understanding of the need for R&D capability in the post-Cold War situation.[92]

President Bush signed Public Law 101-510, the Defense Base Closure and Realignment Act of 1990, on 5 November 1990, establishing the Defense Base Closure and Realignment Commission (DBCRC). A force structure plan and eight selection criteria formed the basis of Defense Secretary Cheney's recommendations to the DBCRC in April 1991. The commission based its review on the Secretary's proposal to realign twenty-nine bases and close forty-three. "The present administration," the Defense Department recommendation noted, "viewed [the] changing world order as an opportunity to implement measured defense reductions. However, Congress has seized upon the reduced threat to our national security and mandated a sharp decline in defense funding."[93]

In response to the FY 1991 Defense Authorization Act, which required each service to recommend a base realignment and closure procedure, the Secretary of the Navy set up the Base Structures Committee. Each service was to make recommendations for closure, which DOD would forward to the DBCRC. The commission's recommendations would have to be accepted or rejected *in toto* by Congress to prevent members from retaining bases important in their particular congressional districts.

In December 1990 the Secretary ordered the Navy's committee to review all Navy and Marine Corps installations and provide a set of criteria, procedures, analyses, and recommendations on realignment and closure of

DA-SC-91-02931

President George Bush inspects the cockpit of a Marine AV-8B Harrier aircraft on board amphibious assault ship *Nassau* (LHA 4) during Operation Desert Shield, Thanksgiving 1990.

shore establishments. In its study entitled "Department of the Navy Base Closure and Realignment Recommendations, Detailed Analysis" (April 1991), the Navy committee used some of the approaches and logic developed during 1989–1990 by the tiger team studies. This report was then forwarded through DOD to the Defense Base Closure and Realignment Commission in accordance with the 1991 authorization act. The Navy's recommendations were calculated to yield annual savings of $568 million for fiscal years 1992 through 1997, although the predicted savings were later reduced by DOD modifications of the plan.[94]

The Navy recommendations called for realignment of Navy RDT&E and engineering capabilities that supported the acquisition process (except Naval Research Laboratory) into four "megacenters": the Naval Air Warfare Center (NAWC), the Naval Command, Control and Ocean Surveillance Center (NCCOSC), the Naval Surface Warfare Center (NSWC), and the Naval Undersea Warfare Center (NUWC). Creating four centers represented implementation of the plan developed earlier by Schiefer and Cann. Ten facilities were recommended for closure, with subsequent consolidation of their functions into other facilities. Another seventeen were recommended for realignment, some with partial relocation to other facilities. In this way, the "BRAC" process was used to implement a long-standing goal of Navy planners, to consolidate the diverse collection of laboratories and R&D centers into a small number of large centrally-directed facilities.[95]

Public Law 101-510 also required the Secretary of Defense to submit to Congress a "force structure" plan assessing probable threats to national security over the period 1992–1997. While details of the plan remained classified, an unclassified summary addressed the end of the Cold War in explicit fashion and provided part of the rationale to the base closures. "America's security agenda is being rewritten because of the collapse of East European communism, the demise of the Warsaw Pact, ongoing changes within the Soviet Union, the reshaping of the U.S.-Soviet relationship, and a reduction in Soviet conventional military power." Regional "quarrels" like the one leading to "armed conflict in the Persian Gulf" and other threats such as insurgency and terrorism would continue. Although the DBCRC's extensive unclassified summary of the force structure report provided the background to its decisions on closure and realignment, the commission made no attempt to link strategic implications of the force structure to specific closures. In short, there was very little effort to demonstrate that planning for R&D was coordinated with planning for future war at this stage.[96]

The Defense Base Closure and Realignment Commission proposed realignment of forty-three bases and closure of thirty-six bases, thereby reducing the Defense Department's "hit list" substantially. In regard to the Navy, the DBCRC recommended closure of seven of the ten RDT&E engineering and fleet support activities that the DOD had recommended closing, and realignment of sixteen of the seventeen other facilities.[97] The commission's estimates would produce recurring annual savings that would total $1.5 billion between 1992 and 1997.[98]

The DBCRC decided in 1991 to retain facilities in Maryland, Virginia, and Washington, D.C., that the Navy was willing to close. Each of the four "megacenters" was created by realigning, closing, or consolidating four to eleven related centers, laboratories, or stations. The thirty-four facilities that combined into the four centers are listed below.

Naval Surface Warfare Center:

1. Naval Surface Warfare Center, Dahlgren, Virginia
2. David Taylor Research Center, Carderock, Maryland
3. Integrated Combat Systems Test Facility, San Diego, California
4. Naval Mine Warfare Engineering Activity, Yorktown, Virginia
5. Naval Coastal Systems Center, Panama City, Florida
6. Naval Ordnance Station, Indian Head, Maryland
7. Naval Ordnance Station, Louisville, Kentucky
8. Naval Weapons Support Center, Crane, Indiana
9. Naval Ship Systems Engineering Station, Philadelphia, Pennsylvania

10. Naval Ship Weapons Systems Engineering Station, Port Hueneme, California
11. Fleet Combat Direction Systems Support Activity, Dam Neck, Virginia

Naval Air Warfare Center:

12. Naval Weapons Center, China Lake, California
13. Naval Air Propulsion Center, Trenton, New Jersey
14. Naval Air Engineering Center, Lakehurst, New Jersey
15. Naval Avionics Center, Indianapolis, Indiana
16. Naval Air Development Center, Warminster, Pennsylvania
17. Pacific Missile Test Center, Point Mugu, California
18. Naval Weapons Evaluation Facility, Albuquerque, New Mexico
19. Naval Air Test Center, Patuxent River, Maryland
20. Naval Ordnance Missile Test Station, White Sands, New Mexico

Naval Undersea Warfare Center:

21. Naval Underwater Systems Center, Newport, Rhode Island
22. Naval Sea Systems Combat Engineering Station, Norfolk, Virginia
23. Naval Undersea Warfare Engineering Station, Keyport, Washington
24. TRIDENT Command and Control Systems Maintenance Activity, Newport, Rhode Island

Naval Command, Control, and Ocean Surveillance Center:

25. Naval Ocean Systems Center, San Diego, California
26. Naval Electronic Systems Engineering Center, Charleston, South Carolina
27. Naval Electronic Systems Engineering Activity, St. Inigoes, Maryland
28. Naval Electronic Systems Engineering Center, Vallejo, California
29. Naval Space Systems Activity, Los Angeles, California
30. Naval Electronic Systems Engineering Center, Washington, D.C.
31. Naval Electronic Systems Engineering Center, Portsmouth, Virginia
32. Naval Electronic Systems Engineering Center, San Diego, California
33. Fleet Combat Direction Systems Support Activity, San Diego, California
34. Naval Electronic Engineering Activity Pacific, Pearl Harbor, Hawaii

As mandated by the law requiring the base closure and consolidation process, the Secretary of Defense commissioned a study by a "Federal Advisory Commission on Consolidation and Conversion of Defense Research and Development Laboratories" to review the steps and recommendations taken so far. The "Adolph Commission," headed by Charles (Pete) Adolph, Acting Director of Defense Research and Engineering, involved six science-administrators from the private sector and six from government laboratories and agencies. Their report addressed why the services should maintain in-house labs rather than contracting out to academia and the commercial sector all of their RDT&E activities. Drawing on the literature of prior studies of this issue, the commission gave ten reasons for maintaining an in-house capacity, mostly derived from the concept that this capacity allowed the government to remain a "smart buyer."[99]

Nevertheless, the report identified many problems with in-house research and development, such as the difficulties in competing for top-level researchers with the private sector, in retaining talented individuals, and in maintaining old and obsolete facilities. It was difficult to obtain funding for the needed combination of capital investment and maintenance required for such facilities.[100]

The Adolph Commission generally endorsed the proposals to consolidate Navy facilities into four major naval warfare centers, but expressed two reservations about the Navy's plans. There was a risk that R&D would lose its separate identity with the merger into full-spectrum centers and there was no high-level official identified to represent and advocate the interests of the centers now that the Office of the DNL was closed. The commission recommended a high-level response to this need. The group also noted that there would be many personnel problems in consolidating facilities and relocating 4,800 people; the turbulence caused by reorganization would lead to loss of key personnel and disruption of work schedules. But, since the Navy's plan did represent a coherent and logical approach to consolidation, the commission recommended that the Navy make every attempt to minimize turbulence and to hold key personnel.

The Adolph group also believed that the Navy's Industrial Funding (NIF) system held promise for the other services, since it required institutions to simulate the methods of private facilities in their search for funding through contracts, program funding, and competitive presentation of proposals to others. By requiring NIF facilities to compete in a market, the Navy achieved some of the benefits of government ownership and control while rewarding initiative and quality as did the private sector.[101]

The Adolph Commission, however, did not address the post-Cold War strategic environment in any detail, but merely sought to define the attributes of a good laboratory and suggest what was being done to achieve those attributes in the emerging system of consolidated Navy centers. In other words, the commission simply regarded base closure and consolidation as an appropriate *budgetary* response to the end of the Cold War. The group did not examine what particular R&D facilities would be most appropriate in dealing with perceived new strategic threats or new types of technical requirements, issues that fell beyond the group's mandate to investigate. In accord with congressional concerns that the services would not close needed facilities, the Adolph Commission focused on maintaining good equipment and good working conditions as the laboratories consolidated. In effect, it applied a checklist to ensure that good facilities survived the closure and realignment process, without making any attempt to determine the relationship of the realigned facilities with future warfare requirements or procurement needs. Its reference to the need for coordination and leadership, however, did result in a Navy response.[102]

The Naval Research Advisory Committee (NRAC) considered the issue of adjustment to the world environment in a 1992 study entitled "Science and Technology (Techbase Strategy for the Year 2010)." Like many NRAC papers and other advisory reports, this document had limited distribution. The NRAC panel advised that the Navy make several long-term adjustments and try to develop a new "paradigm," which would blur the established distinction between Basic Research, Exploratory Development, and Advanced Development. The panel placed emphasis on prototyping methodologies, on simulation and modeling techniques, and on studying commercial manufacturing methods to reduce cost. In addition, NRAC recommended focusing on several specific promising technologies, such as "smart materials" and autonomous vehicles. On the administrative side, the NRAC study recommended more in-house laboratory-industry cooperation, more use of in-house capability to evaluate trade-offs and marginal utility of systems, and the establishment of small study groups of NRL scientists and warfare center specialists to monitor and exploit international developments in areas of science and technology. It echoed the tone and some of the specific approaches of the 1990 Defense Science Board. With its focus on promising technologies and on organizational ways of reducing budget, the study also attempted to address the long-range nature of post-Cold War security threats.[103]

Noting the problems generated by the end of the Cold War, the 1992 NRAC panel stated:

> For more than forty years the United states has invested in a science and technology base geared to support a dynamic arms race with the Soviet Union for superpower military superiority. Now we have won this Cold War, but we are not fully geared to adapt to the victory. Significant realignment of the S&T base must be considered if the United States is to preserve its technological lead and improve its capabilities for the year 2010.[104]

The NRAC panel squarely faced the security issue: "Although the Former Soviet Union no longer constitutes a credible threat to U.S. survival, regional powers will continue to threaten our national interests." These powers would have access to lethal weapons and the threat would be more widely dispersed and unpredictable. The "techbase," NRAC argued, would need to be flexible and creative in its response to this changed threat.[105]

However, the NRAC report suggested that the diverse challenge was compounded by both complacency and declining skills in the Defense Industrial Base. Responding to initiatives by Defense Secretary Cheney, the

1992 NRAC suggested that the older divisions between research and development, which they characterized as the "McNamara Paradigm," needed to be replaced by a "Cheney Paradigm," which would develop a more closely integrated system of requirements-pull and technology-push in the area of development. Like many other advisory reports with limited distribution, this NRAC report did not generate response in the public press despite its profound analysis of fundamental issues regarding the relationship between basic and applied research. Despite its attempt to incorporate a long-range vision, this report did not provide a focused research agenda as did those of an earlier generation, such as the Low, Hartwell, Nobska, and Connolly studies.[106]

In 1993 the NRAC again dealt with policy issues regarding conversion. Noting that the current administration was more concerned with economic impacts than with defense impacts of conversion, the report, in a rather droll fashion, identified a "mission alignment dilemma": "The mission of Defense Conversion, simply stated, is to produce jobs. The mission of the Department of the Navy is to provide an adequate maritime defense. An obvious lack of mission alignment exists."[107]

As in 1992, the NRAC recommended a "paradigm shift," this time to a "bi-directional paradigm" in which technology would transfer freely between the commercial and military sectors. Emphasizing a variety of existing methods by which technology can "spin off" from the military to the civilian sector, and "spin on" to the military sector, the report suggested that with careful planning, defense and Navy technology need not necessarily suffer during the conversion process.

In effect, NRAC pointed the way towards maintaining the material side of the Navy in a high state of readiness and capability, while meeting the publicly emphasized goal of easing the economic impact of defense conversion. Thus, the report stressed collaborative dual-use technologies, partnerships and Cooperative Research and Development Agreements (CRADAs), extension of patent and technology licensing, and, in general, various means of deriving technology benefits for the civilian sector during the conversion process. The report summarized the impact of conversion on defense industries, showing patterns of consolidation (as with General Dynamics) and divestiture (as with General Electric) or, as in the case of some firms, a pattern of simply ignoring the whole issue of conversion.[108]

Early in 1992 the Joint Directors of Laboratories, a group of senior science and technology administrators from the three services and from the Advanced Research Projects Agency (ARPA), produced a study of what had been achieved in the area of tri-service reliance. Although the "White Paper on Tri-Service Reliance" recognized that the emerging new set of strategic requirements, its authors did not attempt to draw an explicit connection between the requirements of the new geopolitical environment and the question of tri-service requirements, even to the extent attempted during the earlier Phase II tiger team study headed by Schiefer. There were two good rationales for tri-service cooperation in the new strategic environment. First, in the reduced funding atmosphere, the services needed to get the most out of the defense dollar. Second, in future littoral and combined operations approaches, close coordination as envisioned in the Goldwater-Nichols Act and as appropriate in special operations would require that the same technologies be used by all three services. In other words, there would be necessarily closer tactical and strategic cooperation among the services; therefore, the technology should be closer.

By mid-1993, Deputy Secretary of Defense William Perry reported to Congress that more than 50 percent of the three services' FY 1993 budgets for science and technology programs (6.1, 6.2, and 6.3a programs) had been "planned for execution" under the Tri-Service Reliance program. As a result, the services had moved from an era where the measure of success was coordination to one in which many efforts were jointly funded or "collocated." He was able to list technological areas in which testing and evaluation would be conducted by one facility operated by one of the services, as the "lead service" for that technology. Throughout 1991 and 1992, the service acquisition executives had entered into "Memoranda of Agreement" setting forth the specific lead service and lead facility for some eighteen different technologies.[109]

The concept that "jointness" would require even closer coordination on the technical side of the so-called C^4I, or command, control, communications, computers, and intelligence, was explored in the operational side of the services in 1993. A joint exercise called Ocean Venture 93 showed the need for "interoperability" of technical equipment if there was to be true joint command of Navy, Marine, Air Force, and Army personnel.[110] However, in the period before such perceptions of the operational side of the Navy and the other services translated into requirements for specific technology, the R&D community tended to think of tri-service reliance not in terms of *products* but in terms of shared or designated research, development, test, and evaluation *facilities*. The "White Paper on Tri-Service Reliance," like the report of the Federal Advisory Commission (Adolph), did not place the issue of tri-service reliance and inter-service cooperation in the context of the new world strategic situation, nor did it attempt to consider the impact of "jointness" or "interoperability" on technology products or weapons. The weapons systems architecture approach was not a factor.[111]

All of the studies that led up to consolidation and those reports reviewing it tended to focus on economic and efficiency issues, with very little effort to address the long-range strategic implications of a post Cold-War geopolitical situation. That gap was soon addressed by the Office of the Chief of Naval Operations.

8

"Limited Intensity Conflict" and Littoral Warfare

During the rise and decline of post-World War II Soviet power and influence, United States armed forces engaged in many small incidents and limited engagements, often with no relationship to the major Cold War issues. When listed, there was a surprising number of such incidents. One study revealed more than 177 engagements between 1945 and 1992, not including actions during the Korean or Vietnam War.[112] These international events demonstrated that the United States needed to continue relying on military power in various forms in the new environment, although at first planners did not explicitly recognize that the small military engagements and peacekeeping actions of the period foreshadowed any future strategic stance. In other words, although U.S. defense forces engaged in great diversity of limited operations, planners continued to regard those actions as exceptions to, rather than typical of, standard defense policy. Schiefer and Cann in 1990 had pointed to that reality as a possible source for doctrinal and weapons-technology demands, but such an approach had received little attention at the time.

In the political environment following 1989, it became clear that the three armed services would be sharing a decreasing budget. Investigators in the Congressional Research Service, in policy analysis think tanks, and in the offices of the Joint Chiefs and the Chief of Naval Operations all sought to generate a restructured set of warfighting goals. To a varying extent, many of these analysts, in what was probably a parallel development of similar ideas, sought to base future military strategy upon the Limited Intensity Warfare that represented the *actual* use to which armed forces had been put in the years following the Paris Peace Accords ending the Vietnam War in January 1972.[113] Under the review of Roles and Functions of the Services mandated by the Goldwater-Nichols Act, the Joint Chiefs and others sought to clarify the strategies and functions that would be appropriate over the coming decades.

NHC L-file/U.S. Army Visual Information Center

Admiral Frank B. Kelso, Chief of Naval Operations, and Marine General Carl E. Mundy Jr., Commandant of the Marine Corps, authors of the naval service's *From the Sea* strategy.

With the publication in September 1992 of the summary statement of the Navy's future in . . . *From the Sea: Preparing the Naval Service for the 21st Century*, written by CNO Admiral Frank Kelso and Marine Corps Commandant General Carl Mundy, naval RDT&E planners had a single explicit and condensed policy upon which to base their expectations. Kelso and Mundy stated that the Navy's role in the future would focus not so much on "blue-water" engagements between major naval forces but upon expeditionary efforts and operations in coastal, or "littoral," waters. This short policy statement reflected the thinking and approach of dozens of articles and reports issued over the prior months by the Center for Naval Analyses and by independent authors.

Reprinted in its entirety in U.S. Naval Institute *Proceedings*, . . . *From the Sea* received widespread distribution throughout the Navy, provoking a wide-ranging and extensive series of commentaries and debates. Kelso and Mundy stated that the experience of the Navy and Marines in crisis response activities suggested that their mission over the next few decades would certainly demand a focus on strategies, tactics, and technologies appropriate to the near-shore environment: "Our strategy has shifted from a focus on a global threat to a focus on regional challenges and opportunities." Thus, hostile actions by third-world powers, often equipped with rela-

DN-SN-91-08058

In the first day of the Operation Desert Storm air campaign against Iraq, a Tomahawk land attack missile (TLAM) is fired from battleship *Missouri* (BB 63).

tively high-technology weapons as well as with older defenses, such as mines and diesel-electric submarines, would require a different set of responses than had the strategies of the Cold War period. One traditional mission, that of sealift, they noted, would continue to be essential.[114]

According to Kelso and Mundy, the "Restructured Naval Force must expand on and capitalize upon its traditional expeditionary roles." That would require a new "mind set" and a new "culture," implying that forces would be swift to respond, structured to project power from the sea, able to sustain long-term operations, and unrestricted by requirements for foreign approval. International respect for freedom of the seas guaranteed naval forces a "unique capability to provide peaceful presence in ambiguous situations before a crisis erupts," unlike land or air transport, which required permission of the intervening nations.[115]

The limited engagements and military crisis responses of recent years, when examined closely, showed a need for continued American preparedness, high readiness, and close cooperation among the services in planning and executing the actions. Should limited engagements characterize warfare in the next decade, they would require continued readiness, continued staffing, and considerable rethinking of military strategy, force structures, and technical requirements. By stressing these once little-emphasized events, military planners would be able to generate a *raison d'etre* for a reconfigured, but flexible and well-prepared, defense force, even though the sort of forces developed to offset Soviet power would no longer be required. For this reason, the burst of research and published reports on limited intensity warfare itself contained an implied policy message. Perhaps most important were the number of limited engagements after 1985. Rather than being exceptions, reports argued, limited engagements had become common and had involved U.S. naval and air assets in particular many times over recent years.[116]

U.S. forces responded to a terrorist bombing in Berlin by a surgical air strike on Tripoli, Libya, in April 1986 (Operation El Dorado Canyon). In 1987–1988, naval vessels in the Persian Gulf provided escorts to Kuwaiti tankers in Operation Earnest Will. The use of naval vessels in drug-traffic interdiction in the mid- and late 1980s suggested that an increased role in policing might be part of the future, representing another, noncombatant type of mission. In December 1989–January 1990, a U.S. force intervened in Panama, arresting General Manuel Noriega in Operation Just Cause and bringing him to the United States for trial on drug charges.[117]

When analysts reviewed Operation Desert Storm, they classified the operation as a "Major Regional Conflict" rather than a "Low Intensity Conflict." Under Operation Desert Storm, launched in response to the Iraqi invasion of Kuwait, U.N. coalition and U.S. aircraft began a sustained bombing attack on Iraqi C^3I and strategic facilities beginning on 16 January 1991. The ground assault of coalition troops from 23–27 February 1991 resulted in Iraqi forces withdrawing from Kuwait.[118]

A subsequent Kurdish uprising in northern Iraq led to the imposition of a "no-fly" zone (Operation Provide Comfort) to protect Kurds who had fled from Iraq government reprisals. Air supply of Bosnian Muslim enclaves during the civil war in the former Yugoslav republic in 1992–1993 (Operation Provide Promise) and participation of U.S. forces during 1992–1993 in reestablishing civil order in Somalia (Operation Restore Hope) demonstrated other roles for U.S. armed forces.

Over the same years, naval and other armed forces were dispatched to emergency and disaster relief operations: Operation Sea Angel, May–June 1991 off Bangladesh; Operation Fiery Vigil in the Philippines, June 1991; and domestic operations in response to Hurricanes Andrew and Iniki, August and September 1992. In the largest noncombatant evacuation operation since the evacuation of Saigon in 1975, naval forces participated in Operation Sharp Edge to transport foreign nationals and diplomats from Monrovia during the civil war in Liberia, May 1990–January 1991.[119] These peacekeeping, humanitarian, and disaster-relief operations gave further examples of the variety of demands and challenges which could be faced by defense forces in the near future.

Operations Prior to . . . *From the Sea* Announcement

Year	Operation	Function
1980	Evening Light	Rescue of hostages (failed)
1986	El Dorado Canyon	Retaliation against Libya for terrorism
1987–88	Earnest Will	Escort of Kuwaiti tankers
1989–90	Just Cause	Arrest of Manuel Noriega
1990–91	Sharp Edge	Evacuation of civilians during civil war in Liberia
1991	Desert Storm	Liberation of Kuwait from Iraqi invasion
1991	Sea Angel	Relief to typhoon victims in Bangladesh
1991	Fiery Vigil	Relief to victims of Mt. Pinitubo volcano eruption
1992–	Provide Comfort	Relief of Kurds in northern Iraq
1992	(Domestic relief)	Relief of hurricane victims in Florida
1992–	Provide Promise	Relief of Muslim enclaves in Bosnia by airdrops
1992–93	Restore Hope	Reestablishment of civil order in Somalia

The ongoing policy debate in the early 1990s over future operational roles of the armed services further complicated the adjustment of the RDT&E community during the period of transitions. Through the years 1992–1993, all of the operations mentioned above were examined in journals of military affairs, and questions of future warfighting configuration of the Navy were repeatedly analyzed, for example, in U.S. Naval Institute *Proceedings* and *Surface Warfare*. The message of all these studies was obvious: combatant forces needed to be kept ready for such uses even if the Cold War ended. Congressional and military staffs and think-tank analysts reviewed the missions of the three services, looking at the precise roles of the Army, Navy, Marines, Air Force, Coast Guard, and National Guard in future warfighting scenarios. The Center for Naval Analyses studied limited intensity conflicts, suggesting that future naval strategy, officer training, and naval acquisitions and R&D should be based on such scenarios.[120]

Admiral Kelso and General Mundy's . . . *From the Sea* made reference to the recent expeditionary activities, including operations in the Persian Gulf and off Liberia, reflecting an effort to base the strategy on actual recent history. The report suggested that the change in approach would affect the very structure of naval administration: "As Naval Forces shift from a Cold

DN-SC-N0724 SCN-91-000064

Carrier sailors load a Tactical Air Reconnaissance Pod System (TARPS) under an F-14 Tomcat on board *Theodore Roosevelt* (CVN 71); it was used in Operation Provide Comfort to photograph Kurdish refugees fleeing Saddam Hussein's forces in northern Iraq, May 1991.

War, open ocean, blue water naval strategy to a regional, littoral, and expeditionary focus, Naval organizations will change."[121]

By 1992 and 1993, the naval RDT&E community, partially restructured following some consolidations and closures, was able to obtain, from OPNAV and public sources, impressions of what military engagements would entail over the next decade. Some planning began to be based on an emerging consensus about the new strategic situation. Although the new emphasis received much attention in the operational, or fleet, side of the Navy, the RDT&E community did not suddenly move to new weapons development responsive to the new thrusts, nor did a technical emphasis spring from any high-level scientific conference like the 1950 Hartwell meeting at MIT or the 1956 Nobska meeting at Woods Hole. Initial attempts to reconfigure electronics programs, to emphasize mine detection and mine countermeasures, and to work on theater ballistic missile defense reflected scattered responses to the new littoral warfare language represented in . . . *From the Sea*. Within military journals, advocates of various technologies and weapon systems suggested their utility in future warfighting scenarios, attempting to line up opinion and support for particular R&D programs.[122]

During the fall of 1992, "working groups, flag-level meetings, and war games" began close examination of technical requirements. By mid-1993, several weapons systems already under development appeared likely to earn further support as . . . *From the Sea* strategy was converted into doctrine and technological requirements. Admiral Kelso identified seven areas of developing systems as particularly crucial to the new strategies. He recommended not to cut from the FY 94 budget: shallow-water antisubmarine technologies; mine warfare and naval gunfire support systems; surveillance systems; systems that facilitate operational maneuver; improvement to standoff weapons; C^4I improvements; and defense enhancements to Aegis ships. Particular systems that received support in this approach were the V-22 tilt rotor craft, electro-thermal guns, and theater ballistic missile defense (TBMD) systems.[123] Again, however, the lack of a central planning agency for naval systems architecture meant that such programmatic recommendations continued to flow piecemeal to the RDT&E community. A structured channel or a high-level science-led analysis of future missions was needed through which "requirements-pull" could become the driver of development.

1992–1993: New Acquisition Strategies, Structures, and Missions

In January 1992 Deputy Secretary of Defense Donald J. Atwood announced at a Defense Department briefing a set of principles that would shape a "New Acquisition Approach." Instead of committing to full production at the point of initiating a program, the DOD would employ a series of approaches constituting a more "Flexible Acquisition Strategy." Atwood suggested that more resources would be devoted to R&D programs that might not go into full production but that would be placed "in the pipeline" or "on the shelf."

A host of specialized procurement strategies or concepts, some new and some developed between 1990 and 1992, were discussed over the next few months, ably summarized and traced to their origins by Professor Charles Cochrane of the Defense Systems Management College in late 1992. More than a dozen flexible acquisition strategies would affect the adjustment of R&D to the changed post-Cold War environment in which full production of new technologies could not be afforded. "Long-shadow," or "Virtual Deployment," involved a commitment only to start on research and development of a weapon system, recognizing that a mere start on the project, such as the Strategic Defense Initiative, might have a deterrent effect on potential adversaries. "Artificial Experience" involved substituting the less expensive computer simulation and modeling for actual test runs in order to move quickly and to reduce environmental impact of actual testing.[124] "R&D Rollover" referred to taking a new system through several iterations, or stages, of development, incorporating lessons learned before final production. "R&D Rollover Plus" required the final production of a working system await either a threat or knowledge that it could alter battlefield operations. "Silver Bullet Procurements" represented rare, highly technological and superior weapons produced in limited quantities, such as the Stealth fighter. "Just-in-Time Weapons" referred to weapons that were capable of production immediately upon need during deployment and mobilization.[125]

In two reports published in 1991 and 1992, the Office of Technology Assessment addressed ways in which private-sector industrial and R&D communities could conform to the constrained defense budget without destroying the nation's defense industrial base. *Redesigning Defense: Planning the Transition to the Future U.S. Defense Industrial Base* and *Building Future Security: Strategies for Restructuring the Defense Technology and Industrial Base* reviewed changes in acquisition procedures and management policies that would allow defense industries to convert to nondefense roles and at the same time be ready for rapid mobilization. The procedures, like those suggested by Atwood, included systems of continuous prototyping and greater reliance on commercial off-the-shelf products.[126]

Such procurement reforms and innovations, under discussion in the R&D community and in the acquisition community more generally, could provide the basis for a new approach to naval RDT&E that would respond to the post-Cold War priorities of a lower defense budget, uncertain threats, sudden deployments or rapid mobilizations, and environmental concerns. However, after the disestablishment of the Office of the Director of Navy Laboratories and the nearly simultaneous creation of four warfare centers, articulating the concerns of the centers at the highest levels of the Navy and coordinating their responses to new requirements needed to be addressed. The need to provide coordination, communication, and control during the chaotic adjustment to the reorganization and to the changing mission became even more pressing.

In response to the concern expressed by the Adolph Commission regarding the gap in RDT&E leadership and coordination, the Navy adopted a concerted plan. Assistant Secretary of the Navy for Research, Development, and Acquisition (ASN[RDA]) Gerald Cann took a three-part approach to the problems of oversight, coordination, and authority over the RDT&E community. For oversight, he established and chaired the Navy Laboratory Center Oversight Committee (NLCOC), with the Vice Chief of Naval Operations and the Assistant Commandant of the Marine Corps as members. The committee met with other senior Navy officials such as the naval systems commanders and various assistant secretaries of the Navy, who attended meetings of the NLCOC in an advisory capacity.

The NLCOC received reports from another group established by Cann to provide coordination, the Navy Laboratory Center Coordinating Group (NLCCG), which consisted of the commanding officers and research directors of the NRL and the four warfare RDT&E centers. The NLCOC-

NLCCG channel thus provided a two-way street through which concerns of the centers and the NRL could be voiced to a very high level, and through which the requirements-pull and other Navy-wide policy decisions could be passed to the centers directly from the assistant secretary level. As an official channel, the NLCOC-NLCCG connection provided both oversight-coordination and a voice for the laboratories and centers, creatively addressing aspects of the administrative gap generated by the disestablishment of the DNL.

Cann established the channel of supervisory authority. Each center, with its focus on one warfare area, could report directly to the concerned system command. In this regard, the new structure resembled the old bureau structure in which the laboratories and development activities were directly administered by one of the bureaus.

Cann's system of oversight, coordination, and authority operated well for a period. However, under his successors in the office of the ASN(RDA), the NLCOC fell into disuse, simply not convening.

Meanwhile, however, the NLCCG, the group of laboratory commanders and research directors, continued to function. Even though the NLCCG no longer had direct access through an established and regular channel to the Navy Secretariat through the ASN, the group continued to coordinate and disseminate information and policy concerns from a variety of sources. The coordinating group developed an expanding agenda of discussion issues and kept the members informed regarding shifting DOD and Navy policies that affected the RDT&E community. In planning reactions to various issues touching on science and technology policy, and in responding to further demands for closure and consolidation, the NLCCG provided mechanisms through which the laboratories and centers could speak with a common voice. Although carrying on many of the coordinating and communication functions of the DNL and of the NLCOC-NLCCG channel, the NLCCG was not in a position to provide the same high-level advocacy voice for the laboratories and centers that some of the individual DNLs had achieved, once the NLCOC atrophied after Cann's departure.

The NLCCG, however, provided its members with current information about new planning and policies that continued to flow from various quarters. NLCCG staff collected and reviewed the numerous reports, studies, directives, and policy documents, passing them to members of the NLCCG for study and discussion.

Follow-up studies in 1992 and early 1993 that acted on Admiral Crowe's suggestion to review the basic functions and force levels of the military included two reports issued by Crowe's successor, General Colin Powell.[127] Endorsed by Secretary Cheney and by President Bush, Powell's *National Military Strategy of the United States* (January 1992) spelled out a "Base Force" that represented a military reduced in strength by proportional cuts across the board but designed to be responsive to crisis situations around the world.[128]

DN-SC-91-07156

General Colin Powell, Chairman of the Joint Chiefs of Staff, during his visit to battleship *Wisconsin* during Operation Desert Shield. In 1992, he produced his own study of defense force levels, *National Military Strategy of the United States*.

A review for the Congressional Research Service, prepared by CRS Senior Specialist in National Defense John Collins, suggested a closer look at future warfighting scenarios and the requirements that such scenarios would impose upon acquisition, training, and allocation of responsibilities among the services. Collins, whose earlier published works examined specialized commando units, gave particular attention to limited intensity conflicts and the roles that various units of the armed services might play in such conflicts in *Roles and Functions of the U.S. Combat Forces: Past, Present, and Prospects* (21 January 1993). Rather than recommending particular cuts or changes in the assignments, Collins suggested alternate courses or options for consideration by Congress for each problem he discussed.[129]

The change in administration from George Bush to Bill Clinton on 20 January 1993 brought even more discontinuity and diversity to the questions of defense materials acquisition, defense R&D, and the future shape of the armed forces structure. Soon after the inauguration of Bill Clinton, the Office of the President issued *Technology for America's Economic Growth* (22 February 1993) over the names of both Clinton and Vice President Al

Gore. The study reflected the new administration's concerns with federal R&D, but contained little reference to future defense needs. Clinton and Gore appeared to be thinking of using the Defense Department not to meet foreign challenges to national security, but to address domestic social and economic issues.

The report visualized the Defense Department as containing resources and facilities that would need to be completely converted to a civilian emphasis and set of goals. That reorientation placed emphasis on environment, jobs, and economic goals. The report did not reflect any concern with the strategic implications of the new international situation. On the whole, the study reflected a frank exposition of the full-conversion philosophy, assuming that with the end of the Cold War there should be a conversion to a peacetime economy. It ignored the contemporary discussion of roles and missions of the services and the emerging doctrines on limited intensity warfare and force projection. In a contemporary fashion, the report echoed the sorts of concerns faced in 1919 and 1946 when the United States dealt with "reconversion" from a war footing to a full peacetime status, focusing on how the resources employed in defense could be redeployed for entirely nondefense purposes. Conversion or reconversion became a driving force behind much of the planning for civilian use of facilities that were closed. The acquisition reforms, which sought to procure commercially produced rather than specially manufactured items with only military applications, also met such a full conversion goal. However, the efforts to purchase "off the shelf" had its limits, as many weapons, platforms, and electronic components did not have analogs or equivalents produced in the nondefense sector.[130]

Implementing an aspect of the conversion philosophy, the Defense Department in 1993 began emphasizing the concept of "dual-use" technologies, an idea that had been stressed years earlier in the Packard Commission Report. Under this concept, the DOD would attempt to purchase as many items as possible in the civilian or commercial sector, not relying upon a separate "defense industry" to provide its needs. Clinton's Under Secretary of Defense John Deutsch stressed to Congress the effort being made to take more advantage of state-of-the-art commercial technologies, to plan further dual-use programs with industry, to cooperate with other agencies and departments of the government in dual-use purchasing, and to evaluate progress in expanding dual-use approaches. The rationale behind this approach was that it would reduce the cost to government of items in common use if defense specifications matched those of the commercial sector.[131]

In September 1993 Secretary of Defense Les Aspin produced a report evaluating the proper force structure of the armed services in response to the emerging world situation. It represented the new administration's rethinking of the Bush "Base Force" proposal of 1992. "The Bottom Up Review: Forces for a New Era" took as a premise the necessity to be prepared to engage simultaneously in two "Major Regional Conflicts."

Presuming a conflict in Korea and in Iraq at the same time as an example, the study reviewed the sorts of force levels and units that would be required.

Conservative commentators criticized "The Bottom Up Review" on several counts. The Heritage Foundation pointed out that the proposed Clinton budgets for the 1993–1997 period, which reduced the defense budget from $270 billion to $246 billion per year, would simply be inadequate for the proposed force levels in the Aspin document, which would require $66 billion more than budgeted over the same four-year period. Critics also challenged the review on the grounds that it would create a "hollow force," one without necessary equipment. In reference to the implications for military R&D, critics asserted that the Aspin plan would lead to a technology gap as older systems become obsolete without replacement.[132] Two weeks after issuing the review, Secretary Aspin reported to Congress that indeed his proposed force levels would require between $13 billion and $31 billion more over a five-year period than had been proposed in the Clinton budget.[133]

Despite criticisms of its costs and force levels, "The Bottom Up Review," like the earlier . . . *From the Sea* and studies of roles and missions generated by Admiral Crowe, Dick Cheney, and others, sought to give a firm direction to the ongoing debates over military readjustment in the post-Cold War world.[134] Such discussions continued into early 1994, with private consulting firms offering seminars on the impact of the littoral strategy on technology.[135]

Over the 1993–1996 period, DOD partially implemented "World Class" procurement, management, and accounting practices. All of these procedures were designed to improve articulation between researchers and end-users, and to make the procurement process less time consuming, more efficient, and less costly. The approach was essentially driven by managerial concerns, although Paul G. Kaminski, Under Secretary of Defense for Acquisition and Technology, addressed both managerial and strategic issues in a number of speeches and in congressional testimony in early 1996.[136] Reforms and methods at the DOD level, however, did not produce a clear picture in the Navy R&D community of the sorts of technological requirements that would be generated by future missions.

9

The Chaos of Policy Formulation

Policy formulation in the Department of Defense, as in other branches and agencies of the U.S. Government, does not proceed in a clear, simple direction from any focal point of decision making. Rather, like other policies, defense policy emerges from extended discussions and exchanges among the executive and legislative branches, through their staffs, and from associated consulting groups and think tanks. When decision makers are faced with a changing world, the required shift in policy gradually evolves from the ongoing debate. Such a complex process occurred in 1947 when the Navy adjusted and developed a Maritime Doctrine for the Cold War.[137]

Doctrinal readjustment to the post-Cold War environment and its impact on RDT&E priorities produced a chaotic unfolding of positions and a similarly intricate interplay of advocates of different scenarios in the late 1980s and early 1990s. Some in the naval RDT&E community lamented the lost opportunity to develop a systems architecture response within SPAWAR. Without a single office on the Navy's material side responsible for developing a response to new warfare scenarios, other offices were free to generate quite a variety of weapons acquisition policies.

The various flexible acquisition policies articulated by the Office of the Secretary of Defense between 1992 and 1996 suggested that new approaches would adapt to the post-Cold War environment. Although many of those approaches had implications for naval RDT&E, they provided more of a menu of options rather than a centrally declared policy or a planned set of procedures for the Navy. Perhaps the very nature of the DOD relationship to the services, in which the services retain a degree of autonomy over planning and over their own R&D emphases, has contributed to the apparent lack of central direction both after World War II and in the early 1990s. In each situation, competing views on the future of warfare vied for attention.

On the operations side of the Navy, the CNO considered the demands of the post-Cold War era in light of providing direction for the fleet. Despite

Admiral Kelso's and General Mundy's attempt in 1992 to capture and define a policy change in . . . *From the Sea*, their statement represented not so much a setting of policy but rather an expression of one position among many. Since under Goldwater-Nichols the ultimate responsibility for missions and goals fell with the Secretary of Defense, Aspin's "Bottom Up Review" tended to have more impact than did the CNO's focus on limited intensity conflicts even though many planners in Navy organizations and analysts publishing in naval opinion journals followed up on Kelso and Mundy's lead.

During the period of transitions from the mid 1980s through 1996, some of the major policies that bore on naval RDT&E facilities reflected contradictory positions. In the early years of the period, observers like those writing the Packard Commission Report and the Coopers & Lybrand report stressed the need to *modernize facilities and make them efficient to meet the Soviet threat*. Certain aspects of the Reagan administration's preference for the private sector over the government influenced the thrust of later reports. These reports argued that the Navy should make the RDT&E establishment *more reliant on the private sector to make that establishment more responsive* or, on the contrary, that the government should strive to make the facilities *more competitive to meet the challenge of international trade and international competition in the areas of science and technology*.

Responding to the intent of the Goldwater-Nichols Defense Reorganization Act and recommendations in the Packard Report, Secretary of Defense Cheney sought to make the services' facilities *more inter-service reliant*. In facing the question of base closing, the Schiefer report noted that facilities should not only respond to reform pressures, but be able to *respond to limited intensity conflict scenarios*. In general, the Defense Department's base closure commission sought to make the facilities *more streamlined to meet "peace dividend" budget needs*, and *more capable of cradle-to-grave service*. Through 1992 and 1993, some analysts argued for *bringing tri-service reliance and "jointness" to bear upon the technological systems* that supported the command structure. Some staff in the new Clinton administration argued that the defense establishment should *convert entirely to civilian objectives* and refocus the DOD on priorities such as employment, regional economics, and environmental considerations.

The CNO and the Commandant of the Marine Corps provided strong arguments for *planning around limited intensity warfare approaches and strategies* in . . . *From the Sea*. Secretary Aspin argued in the "Bottom Up Review" for *maintaining a force structure sufficient for two major regional conflicts*, although critics claimed his projections were inadequate to that purpose. Procurement emphasized *dual-use technologies* as a means of reducing costs and strengthening the U.S. competitive position in the world economy.

For the most part reports and studies reviewed in this essay adhered to only one of the positions listed above, although the 1990 tiger team studies and a memorandum by Assistant Secretary Cann recognized the interplay of several issues. By 1993 and 1994 some planners sought to translate the imperatives of limited intensity conflict into late twentieth-century technology and to develop a coherent force structure and tactical and doctrinal programs responsive to the needs of those warfare scenarios. They took that course with little support from the White House or from the Secretary of Defense. The challenge to management of full-spectrum RDT&E facilities would be to adapt to the product demands characteristic of the post-Cold War international setting. As in prior decades, keeping the facilities responsive to the needs of the fleet would remain a high priority. That responsiveness would be especially essential as these needs underwent sweeping reassessments.

Despite Mikhail Gorbachev early reforms in the Soviet system, the future was impossible to foretell in 1986. Even as Eastern European developments moved rapidly and as U.S. forces engaged in low-level conflicts around the world during the next three years, few analysts could develop sufficient distance or perspective to produce new doctrinal or strategic thinking, and instead continued to focus on DOD inefficiency and poor procurement practices. The perceptive 1990 Navy tiger team study did anticipate some of the developments quite accurately, as did some reports from the Center for Naval Analyses. With the publication of . . . *From the Sea* in 1992, a degree of guidance through the uncharted waters of the post-Cold War environment became available from the CNO and from OPNAV planning offices. After the election of the new administration in 1992, defense journals and planners within OPNAV entered a thorough debate over limited intensity warfare, the meaning of the end of the Cold War, and the sorts of technology appropriate in the coming years. For Navy policymakers, the years 1990–1992 included the end of the Cold War and the uncertain beginnings of a post-Cold War world.

However, as Thomas C. Hone pointed out in his history of the Office of the Chief of Naval Operations, *Power and Change*, the "various offices within OPNAV compete with one another, and often there is no logical connection between force and strategic planning and programming."[138] The effort to coordinate and connect the fleet's view of the Navy's future with the planning for RDT&E would clearly continue as a difficult challenge through the 1990s. Current weapon systems would adapt and new ones would be developed, but without a coordinator to ensure that they were combined in a single architecture and without a leader to ensure that strategic plans became translated into material acquisition programs. The structure established by ASN(RDA) Cann provided a system for overseeing, coordinating, and supervising the NRL and the four warfare centers. Despite the failure to maintain the channel of communication by holding regular meetings of the NLCOC, the NLCCG continued to collect, gather, and sort out the rapidly evolving policy materials through the mid-1990s to keep on an intelligent and responsive path through the chaos.

10

The Post-Cold War Public Discussion

The public policy literature of the 1980s and early 1990s provides some understanding of the difficulties encountered by defense firms and government defense facilities as they faced conversion in the post-Cold War period. As the Cold War showed signs of ending, dozens of studies and commentaries published in news magazines, business periodicals, and scholarly journals analyzed the economic and business impact of converting the U.S. defense industry to the manufacture of commercial and civilian products. The studies ranged from superficial treatments based on one or two interviews to more extensive reports on conferences, econometric studies, and full-scale monographs and books. Although any bibliography of widely published studies on defense conversion would contain more than a hundred items, a review of the literature shows that only a small number of these articles and reports raised issues of strategic or tactical preparedness.

Looking at twenty-five years of efforts to convert defense industries to civilian pursuits, Gordon Adams, director of the Defense Budget Project in Washington, D.C., claimed in the mid-1980s—at the height of the Reagan administration's defense buildup—that "economic conversion has come to a dead end." According to Adams, the conversion efforts in the 1960s and 1970s had "borne little fruit." Activists and legislators, he said, continued to "beat their heads against an impossible, utopian wall." Giving details of alternate-use plans proposed for General Dynamics and Douglas Aircraft, and from plans proposed in Great Britain and by U.S. labor unions, he traced, one by one, how each had to be abandoned. Adams viewed the conversion plans as political efforts to widen the constituency for arms control by holding out the hope of a peace dividend, rather than as genuine efforts to do serious economic planning on a local and national level.[139]

As signs of the Cold War ending became more clear in the late 1980s, other business observers and economists continued to call for centralized planning to deal with conversion of military to civilian production. Anticipating the benefits that planning would have for international relations, the economy, and industry as a whole, Seymour Melman, author of two books on conversion from military to civilian production, argued in a 1986 essay in *Technology Review* that careful conversion planning would be essential in slowing down the arms race and could end the decline of U.S. industry. Melman reviewed the case of Boeing-Vertol. When that company tried to convert from making military helicopters to producing electric powered trolleys and subway cars in the 1970s, it "met with disastrous results." Having dispensed with prototype testing, the firm put the electric trolleys directly into service with the Massachusetts Bay Transportation Authority, where they frequently broke down. After a series of lawsuits, Boeing-Vertol went back to manufacturing military helicopters. Melman argued that the case demonstrated the need for careful planning by corporate officials facing conversion tasks.[140]

Not all of the early commentators were optimistic about a potential peace dividend; some anticipated it would take several years before positive economic effects would be felt. Some dryly admitted the short-run costs of a cut in defense spending "could be pronounced." As companies competed for contracts or converted to new business, workers would find the "peace dividend might be more like a peace dole."[141]

Nevertheless, industrialists could be found making upbeat comments in the popular press about the prospects of a peace dividend and its economic and business benefits. Some businessmen were "surprisingly optimistic about what they see as an opportunity to diversify and free themselves from the bureaucratic millstone that has held them back in the international commercial market." Some analysts anticipated a payoff at the Pentagon, because the trend to eliminate the barriers between the defense industry and the commercial market could lead to cheaper and possibly more efficient weapons.[142]

Shaping Post-Cold War Conversion—The Economic Focus

In 1988 Mikhail Gorbachev announced the planned withdrawal of 500,000 Soviet troops from the Warsaw Pact nations of Eastern Europe. By late 1988 and early 1989, the press, particularly *Newsweek* and the *Washington Post*, was announcing a forthcoming "peace dividend." In December 1989 the Berlin Wall was taken down, and in the early months of 1990, a series of elections across Eastern Europe installed noncommunist governments. In November 1990 President Bush and Party Secretary Gorbachev jointly announced that the Cold War was over. Disarmament and downsizing had already begun in the United States in response to the decreased threat from the Eastern bloc.[143]

Rather suddenly, in 1990–1991, the Navy was cut back. Four of the Navy's aircraft programs had been eliminated. In 1991 the P-7A antisubmarine warfare aircraft and the A-12 Avenger were both canceled, and the planned F-14D Tomcat remanufacturing program and the Navy version of the Advanced Tactical Fighter were cut out of the Navy's 1992 budget. The planned 600-ship Navy never reached more than 559 ships; in 1992 the goal was reduced to 476 and then cut further from there. Admiral Frank Kelso announced plans to aim for a 450-ship Navy. The Navy cut back its order for three or four *Seawolf* class submarines a year to one per year and eventually closed one of the two yards capable of building the submarine. In the 1980s the Navy had purchased, on the average, nine combatants and eight amphibious, or support, ships per year. In the 1990s the plan was scaled down to purchase, at the maximum, five combatants and four amphibious/support ships.[144] In 1991 base closure and consolidation reorganized the thirty-four naval RDT&E facilities into four megacenters, or warfare centers; further changes in the 1993 and 1995 cycles of base closures and consolidations preserved the four warfare centers but shut down some of the smaller locations associated with the centers.

At the beginning of the 1990s, a few commentators drew parallels to the end of World War II. In a study included as an annex to the 1993 Defense Conversion Study, William G. Stewart provided a close historic review of the economic conversion policies adopted during and after World War II, during the Korean War, and after the Vietnam War. However, this analysis focused on questions of economic policy and planning rather than on the impact of defense conversion and reduction on national security, readiness, and the capability of the defense industrial base for mobilization.[145]

As a young man, economist Herbert Stein wrote a prize-winning essay on post-World War II conversion in 1945, doubting whether the economic impact of conversion would be disastrous in the post-Cold War period. He noted that at the end of World War II, unemployment rose from a low of 1.2 percent in 1944 to less than 4 percent between 1946 and 1948, not a disaster. Similarly, production of consumer durables climbed dramatically in the 1940s. Because of that experience, Stein commented, "I have never been able to take very seriously worries about the bottleneck and maladjustment problems associated with smaller conversions in the fifties [post-Korea] and seventies [post-Vietnam] and perhaps now in the nineties."[146]

Some commentators went further in their optimism, hailing the end of the Cold War as worthy of celebration because it held the potential to reshape U.S. policies. Liberal critics believed that calling the money to be saved from defense a "peace dividend" tended to minimize the massive impact visualized. One claimed that over the decade of the 1990s, some $1.5 trillion could be saved, and "social action groups" would no longer have to struggle over a "diminishing pie." Rather, they could "coalesce to claim the resources necessary to rebuild this country from the bottom up and the inside out."[147]

Despite such eager optimism, a few economists predicted a slump before any positive effects of the peace dividend could be achieved. Working with modeling and a scenario that incorporated a 25-percent reduction in U.S. defense spending, one economist concluded that "the contractionary fiscal policy in both the United States and other developed economies initially overwhelms any positive effect on investment which the defense cuts may have. . . . Clearly, some industries benefit more than others, with defense related manufacturing industries suffering the most."[148] Such dire, if dry, predictions were borne out over the next three years.

In 1990 the Congressional Budget Office took note of these commentaries from economists and warned that defense conversion would entail difficulties.

> These favorable security developments have raised the prospect of large cuts in the U.S. defense budget, which some have labeled the "peace dividend." While almost everyone favors reallocating resources that no longer need to be spent on defense, concerns have been raised about problems of economic dislocation and management that could occur as the country makes the transition to lower defense budgets.[149]

In 1991 President Bush optimistically anticipated that the conversion after the Cold War would be on a much smaller scale than post-World War II conversion, following the logic of economist Herbert Stein. In the 1991 *Economic Report of the President*, Bush noted:

> Proposed cuts in defense spending over the next few years start from a much smaller share of GNP and are modest in size relative to the demobilizations after World War II, the Korean war, and the Vietnam war. The economy will adjust smoothly to reductions in defense spending, but some workers and firms will need to adapt to new circumstances. Programs are in place to help workers and communities adjust to reductions in defense employment.[150]

Yet Murray Weidenbaum, formerly of the Council of Economic Advisors and currently director of the Center for the Study of American Business at Washington University in St. Louis, suggested in his 1992 book *Small Wars, Big Defense* that cultural barriers in the defense industries would prevent conversion from reaching its potential for benefits.

> It is not hard to understand why defense company managements have become so reluctant to move from fields they have mastered into lines of business alien to them. Their lack of knowledge of nondefense industries is pervasive. It includes ignorance of products, production methods, advertising and distribution, financial arrangements, contracting forms, and the very nature of the private customer's demands.[151]

In another piece reflecting the same themes, Weidenbaum drew from extensive literature on the economic impact of conversion, showing that defense industries, in order to meet the needs of defense procurement, developed internal procedures and practices that were quite different from those of industries serving purely nondefense, commercial markets. Defense industries had low capital, little marketing capability, and very little experience in producing in high volume at low unit cost. They shaped their reporting structures to meet the needs of government clients.[152]

Several business and economic observers echoed Weidenbaum's arguments, pointing out a variety of reasons why defense conversion was difficult for most defense industry managers.

> Defense contractors in the United States face a painful choice between downsizing or investing in new high-risk commercial ventures. Past experience reveals numerous failed efforts to penetrate commercial markets and few, if any, successes.[153]

Bruce Berkowitz, a national security analyst at Carnegie-Mellon University, held that the practices in the industrial sector would be hard to change and that many defense industries would not make the transition comfortably. He noted that the defense community had been "notoriously ineffective and inefficient" in transferring technology. Berkowitz attributed these problems not to the technology itself but to the "deep seated organizational and cultural obstacles within the military which discouraged and impeded technology transfer." Furthermore, DOD acquisition policies made it "virtually impossible," he claimed, for products to be cost-competitive in commercial markets.[154]

Defense cuts hit some specific regions, occupations, and industries much harder than others. Defense industry workers were not easily transferable to alternative employment. A detailed 1992 study published in the *Journal of Economic Perspectives* traced the projected impact of employment cuts in defense industries for the 1991–1996 period by state, showing that the heaviest impact would be in the District of Columbia, Alaska, Hawaii, and Virginia in terms of percentage of overall work force. The highest total DOD employment cut in sheer numbers would be in California, with over 180,000 jobs expected to be cut in the period.[155]

By 1992 some commentators were growing even more pessimistic about the economic impact. The problems confronting the United States during defense conversion were common to other states in both the Western bloc and the former Soviet bloc. One specialist in international economics noted that euphoria had given way to the concrete problems of dismantling much of the vast defense arsenal built up over fifty years. The countries hardest hit were the United States and the republics of the former Soviet Union, but the problem had global ramifications. There was "no clear idea of what a post-Cold War security framework" should be like. New regional economic groupings would "emerge from the wreckage of Cold War political alignments."[156]

In the United States, without central planning for changed defense requirements, the haphazard drawdown of defense budget threatened to destroy the industrial base, with some security implications. Loral Corporation CEO Bernard L. Schwartz, writing for the Foreign Policy Institute at Georgetown University, noted that the lack of a firm sense of security requirements on which to base budget could "cause a meltdown of this country's industrial base and therefore our ability to defend our nation."[157]

Epilogue

Rather than yielding simple general patterns, this initial broad survey of prior demobilizations and the Navy approach to base closure reveals an intricate interplay between fleet needs and shore capabilities, between delayed and implemented innovation, between hardware and doctrine, and between the international context and the domestic budget. During peacetime, the Navy sought to preserve a shore establishment with sufficient programmatic strength and personnel to provide a backup to the fleet and to address continuing needs to modernize. Some strong centers and facilities were maintained and even expanded during most peaceful interwar periods. In the 1950s, several crucial planning meetings, especially the Nobska meeting, gave shape to the Navy's undersea program, while the far-seeing Connolly study spurred on missile development.

The maintenance of core capabilities in terms of skills and teams varied from facility to facility, from period to period, and from program to program. Work on both major innovative systems and ordinary hardware sometimes faced neglect and delay, and at other times kept the United States minimally ready for the next conflict. The material side of the Navy, while usually de-emphasized during peacetime, was never entirely neglected. Emphasis on particular programs in the interwar periods tended to reflect extended work on major innovations from the last war, as with the submarine and aircraft between World Wars I and II, and with the nuclear weapon, the missile, and the nuclear reactor as a source of propulsion, and with the implications of the modern true submarine, during the 1950s and 1960s.

In the 1990s, no clear picture immediately emerged from the chaotic interplay of visions of the future. Admiral Kelso's effort to use the lessons of the recent past to visualize the near future was the beginning upon which much of the planning in the mid 1990s was based. In 1956 the scientists and planners at Project Nobska visualized the future submarine threat and stimulated a host of new research directions. In the late 1990s, a similar conference might choose to focus on the weapons and systems which the U.S. Navy will be likely to confront in the early 21st century. To date, however, the primary driving force behind planning and reorganization continued to be a search for efficiency and economy, rather than a response to future defense needs.

Some budget planners, during the prior and more recent periods of cutbacks, have considered cutting the "tail" instead of the "tooth," stripping down the rear echelons at home before reducing the ships and men of the fleet. Generally, however, during any postwar conversion, both the fleet and the shore have preserved a core that provided continuity and a basis for later expansion. The NRL and the warfare centers, which together inherited the Navy's long tradition of RDT&E, worked to keep abreast of the rapid changes and to maintain the strengths of that tradition.

The lessons of history were there, and the Navy appeared to be heeding them: the tail would have to be kept strong and healthy, or the tooth would partially decay.

Chronology

International Events	Defense Department Events and Report
1985	
March Gorbachev elected General Secretary General, Communist Party, USSR	
1986	
April Operation Diablo Canyon: U.S. air raid on Tripoli, Libya, in response to Berlin terrorist bombing	
	June Coopers & Lybrand, "Management Analysis of the Navy Industrial Fund Program: Naval Laboratories Review Report"
	Goldwater Nichols Department of Defense Reorganization Act, PL99-433 signed
October Reykjavik Summit Meeting	*October* Anti Drug Abuse Act signed authorizing DOD participation in drug law enforcement
1987	
July Operation Earnest Will: U.S. naval escort of Kuwaiti tankers	

December
Intermediate-Range Nuclear Forces (INF) Treaty signed by Gorbachev and Reagan at Washington Summit Meeting

December
DSB 1987 Summer Study

1988

April
Operation Praying Mantis: punitive strike on Iran offshore oil platforms

Soviet withdrawal from Afghanistan

October
DSB 1988 Summer Study

November
George Bush elected U.S. President

December
Cut of 500,000 Soviet troops in Eastern Europe announced by Gorbachev

End of Cold War announced by Prime Minister Thatcher

1989

Multiparty Election, USSR

April
Office of Technology Assessment, "Holding the Edge: Maintaining the Defense Technology Base"

June
Democratic vote, Poland

June
SECDEF Cheney Report on Defense Reorganization

July
End of Soviet control in Eastern Europe

July
Department of Navy Management Task Force Study

August–December
Hungarian border opening; East German exodus to West Germany via Hungary

September
JCS Chairman ADM Crowe statement on Roles and Missions

October
Office of SECDEF, "DOD Management of Technology Development—Implementation of Recommendations of Defense Management Review"

November
Press reports of Peace Dividend

December
Destruction of the Berlin Wall

December
DMRD-922 recommending consolidation of RDT&E facilities issued

1990

January
Operation Just Cause: U.S. capture of General Manuel Noriega

February–May
Tiger team studies forwarded through channels

March
Democratic election, Hungary

April
Democratic election, Romania

May
Operation Sharp Edge: Civil War and evacuation of civilians from Liberia

Democratic election, East Germany; unification with West Germany approved

June
Democratic election, Bulgaria

Democratic election, Czechoslovakia

July
Resignation of Boris Yeltsin from Communist Party

July
Director of Navy Laboratories, "A Review of Studies Conducted from November 1989 to April 1990 on Restructuring the Navy RDT&E Community."

August
Iraq invasion of Kuwait; post-Cold War world order announced by President Bush

October
German reunification

November
End of Cold War jointly announced by Bush and Gorbachev; Conventional Forces in Europe (CFE) Treaty signed by both

1991

January
Operation Desert Storm: U.S. and allied bombing phase over Iraq

Operation Eastern Exit: evacuation of civilians from Somalia

February
Operation Desert Storm: ground war phase in Iraq

March
Operation Southern Watch: air interdiction, Southern Iraq

April
Operation Provide Comfort: air interdiction Northern Iraq

April
Department of the Navy Base Structure Committee, "Department of the Navy Base Closure and Realignment Recommendations, Detailed Analysis."

June
Yeltsin elected President of Russia; dissolution of Warsaw Pact

July
START I Treaty, scheduling mutual reduction of strategic nuclear weapons, signed by Bush and Gorbachev in Moscow

July
Defense Base Closure and Realignment Commission Report

August
Attempted coup and detention of Gorbachev in USSR

September
Independence from USSR declared by Latvia, Lithuania, and Estonia

September
"Federal Advisory Commission (Adolph Commission) on Consolidation and Conversion of Defense Research and Development Laboratories—Report to the Secretary of Defense"

December
Dissolution of USSR by constituent republics that form Commonwealth of Independent States(CIS)

Resignation of Gorbachev

1992

January
Office of the DNL closed

Joint Directors of Laboratories, "White Paper on Tri-Service Reliance"

New DOD Acquisition Approaches announced

JCS Chairman, *National Military Strategy of the United States*

May
Lisbon Protocol signed by Russia, Belarus, Ukraine, and Kazakstan to abide by START I

July
Operation Provide Promise: air relief to Bosnian enclaves

September
General Accounting Office, *United Nations: U.S. Participation in Peacekeeping Operations*

September
Chief of Naval Operations and Commandant of the Marine Corps, *. . . From the Sea*

November
Bill Clinton elected U.S. President

November
Naval Research Advisory Committee, "Science and Technology—Techbase Strategy for the Year 2010"

1993

January
START II Treaty signed by Presidents Bush and Yeltsin

Clinton Inauguration

Shelling of Russian Parliament

September
SECDEF Aspin, "Bottom-up Review"

December
Russian Parliamentary elections

Endnotes

1. There is an extensive body of literature on U.S. entry into World War I. A clear and concise coverage can be found in Ernest R. May, *The World War and American Isolation, 1914–1917* (Chicago, 1966).

2. Rodney Carlisle, *Powder and Propellants: Energetic Materials at Indian Head, Maryland 1890–1990* (Indian Head, 1992), 55–59.

3. Ibid., 68–70.

4. The superiority of Japanese torpedoes and aircraft is detailed in Gordon Prange, *Miracle at Midway* (New York, 1986); details regarding the *Zero* are reported in Robert L. Caleo, "From Autos to Aircraft: General Motors' WWII Conversion to Wildcats and Avengers," *Naval Aviation News*, July–August 1995, 26–32.

5. Allan R. Millet and Williamson Murray, "Innovation in the Interwar Period," Contract Report MDA903-89-K-0194, Office of Net Assessment, June 1994, iii.

6. The examples described above are treated in Millett and Murray, "Innovation in the Interwar Period," 145–214 (strategic bombing), 215–300 (close air support), and 433–492 (radar).

7. Rodney Carlisle, "Where the Fleet Begins," History of David Taylor Research Center, work in progress.

8. James Phinney Baxter III, *Scientists Against Time* (Boston, 1946).

9. Gordon Prange, *Miracle at Midway*. Prange discusses a wide variety of failures of U.S. aircraft and ordnance: 173, 175, 195, 198, 204, 208–12, 252, 258, 268.

10. Walter Dornberger, *V-2* (New York, 1954), 107.

11. Thomas Powers, *Heisenberg's War* (New York, 1993), regarding German failure to develop the nuclear weapon, accepts to an extent the German scientists' own claims that they made a conscious decision not to pursue the weapon. In an earlier work, *German National Socialism and the Quest for Nuclear Power, 1939–1949* (Cambridge New York, 1989), Mark Walker suggests that the failure to develop the weapon came from an interest in developing a power reactor from nuclear energy, a project that German scientists decided to postpone until after the war.

12. See C. D. Bekkar, *Defeat at Sea* (New York, 1955), 75–88, for a popularized treatment of the issue of superior Allied radar. More scholarly reviews of the problem are in Tony Devereaux, *Messenger Gods of Battle: Radio, Radar, Sonar: The Story of Electronics in War* (London, 1991). The German nuclear program is covered in Powers, *Heisenberg's War*.

13. Harvey Sapolsky, *Science and the Navy: The History of the Office of Naval Research* (Princeton, NJ, 1990), 57, 72, 109. Although begun under a naval contract about 1980, this work was not completed until 1988 and not published until 1990. The comparisons and observations by the author do not incorporate a recognition of the end of the Cold War.

14. "Study of Undersea Warfare" (Low Report) (22 April 1950), 26, 27, 62, 65, 66 (declassified copy at Naval Historical Center).

15. A thorough treatment of the Low Report and the Hartwell Project can be found in Gary Weir, *Forged in War* (Washington, 1993), 133–47.

16. Tamara Melia, *Damn the Torpedoes* (Washington, 1992), 67.

17. Ibid., 77–79.

18. Weir, 218, and Chapter 12. A declassified version of Volume I of the final report on file at the Naval Historical Center provides detailed reports and recommendations of the various groups: "Project Nobska—The implications of Advanced Design on Undersea Warfare. Volume I: Assumptions, Conclusions and Recommendations" (Washington, 1956), 25, 31, 32, 48–49.

19. The information regarding nuclear weapon delivery systems is documented in Chuck Hansen, *U.S. Nuclear Weapons: The Secret History* (New York, 1988), 203–9.

20. R. E. Kistler and R. M. Glen, *Notable Achievements of the Naval Weapons Center*, NWC TP 7088 (China Lake, Naval Weapons Center, 1990).

21. Research Office, Office of the Under Secretary of Defense for Research and Engineering, "Required In-House Capabilities for Department of Defense Research, Development, Test and Evaluation" (Perry Report) (1 October 1980), 6–7.

22. Edward Marolda, *The United States Navy and the Vietnam Conflict*, Vol. 2, *From Military Assistance to Combat, 1959–1965* (Washington, 1986), 277–97.

23. Details regarding both the Walleye and the Snakeye can be found in Kistler and Glen, *Notable Achievements of the Naval Weapons Center*.

24. Ibid., 31, 35.

25. Melia, *Damn the Torpedoes*, 97.

26. Carlisle, *Powder and Propellants*, 175. For World War II production, see 107–11; for quantity production, see Kistler and Glen, *Notable Achievements of the Naval Weapons Center*, 21.

27. Carlisle, *Powder and Propellants*, 175, 179, 236.

28. Rodney Carlisle, *Management of the U.S. Navy Research and Development Centers During the Cold War Era: A Survey Guide to Reports* (Washington, 1996), 51–62.

29. *Washington Post*, 18 November 1988, 1.

30. Journalists anticipated defense budget cuts as early as December 1988 (*Newsweek*, 19 December 1988, 30). Discussion of "peace dividend" as a phrase was reported and evaluated in a *Washington Post* editorial, 27 November 1989.

31. Office of Technology Assessment, *After the Cold War: Living with Lower Defense Spending*, OTA-ITE-524 (Washington, 1992). Suggestion that the Cold War ended in 1991 appears in the preface, iii, although its figures demonstrate reduced national defense spending, reduced total defense employment levels, and reduced defense spending as a percentage of Gross National Product in 1990 and 1991 (4–5).

32. White House Science Council, "Report of the White House Science Council, Federal Laboratory Review Panel," (Packard Report) (May 1983), 8.

33. Ibid., 9n, 9–10.

34. Ibid., 11–12.

35. See especially Dick Cheney, "Defense Management: A Report to the President" (July 1989).

36. John E. Morton noted the number of CRADAs in "The U.S. Navy in Review," U.S. Naval Institute *Proceedings* (May 1993): 125.

37. Dr. James Colvard, interview with Rodney Carlisle, 13 May 1994.

38. For disestablishment of NAVMAT as part of the move to reduce bureaucracy, see interview, Joe Marchese with Admiral James B. Busey, former Vice Chief of Naval Operations, 20 April 1990, 4.

39. Busey interview, 7. The transfer of the architecture and systems role had been recommended by Dr. James Colvard in a brief memorandum that he issued in retrospect as former Deputy Chief of Naval Material, the "Haul Down Report" (10 December 1985).

40. Robert Doak, Deputy Commander of SPAWAR under Vice Admiral Glenwood Clark, attributed the difficulty in defining the architecture mission to the problem of getting the new office structured. Robert Doak, interview with Joe Marchese, 17 April 1990, 4.

41. Rear Admiral Wayne Meyer believed that as the first commander of SPAWAR, Vice Admiral Clark was restricted by his experience in the Special Projects Office, that he did not have the necessary breadth of experience to take on the architecture role, and that he remained unsupported by Secretary Lehman, who had created the post. Meyer, interview with Joe Marchese, 30 April 1990, and 9 May 1990, 31–45.

42. Howard Law, who had worked in the office of the Director of Naval Laboratories through the tenures of several DNLs, saw the weakening of that office and its elimination as a severe blow to the central source of leadership. Law, interview with Joe Marchese, 6 June 1991, 45–47. This view was also held by Dr. James Colvard; see Colvard interview.

43. Dr. James Colvard believed the original appointment of Vice Admiral Clark to the command of SPAWAR was too junior an appointment for the responsibilities of the task, and that the electronics responsibilities would interfere with the system-wide mission. Colvard, interview with Joe Marchese, 15 December 1990, 8, 12.

44. Robert Scarborough attributed part of the problem to the practice of rotation of military officers, and part to the persistence of cultural patterns from the Special Projects Office. Scarborough and Steven Coffey of Booz, Allen & Hamilton, interview with Joe Marchese, 9–17 May 1990, 24, 50.

45. Rear Admiral Craig Dorman held a doctorate in Oceanography and worked in antisubmarine warfare in SPAWAR. He attributed the difficulties of SPAWAR to the persistence of multiple planning functions in other offices. Dorman, interview with Joe Marchese, 6 March 1990, 13.

46. Gerald Schiefer noted the problem of compartmentalization, among others. Schiefer, interview with Joe Marchese, 1 November 1991, 31–32.

47. Meyer interviews, 30 April and 9 May 1990.

48. Jon P. Walman, "Expeditionary Warfare—New Emphasis for Naval Operations," *Surface Warfare Magazine*, November–December 1993.

49. Colvard interview, 13 May 1994.

50. Coopers & Lybrand, "Management Analysis of the Navy Industrial Fund Program: Naval Laboratories Review Report" (June 1986), Intro-1, Org 7-8, Intro-2.

51. Ibid., Chronology, Appendix 1.

52. Ibid., Org-1, Org-7, Per-1, Pro-1, Fin-1, Fin-3, Mis-2, Mis-4.

53. Defense Science Board, 1987 Summer Study, "Technology Base Management" (December 1987), ii.

54. Ibid., ii–iii.

55. Ibid., v, 2, 3.

56. Defense Science Board, 1988 Summer Study, "The Defense Industrial and Technology Base" (October 1988), 1, Executive Summary.

57. Ibid., 1, Executive Summary.

58. Ibid.

59. Ibid., 2–8, Executive Summary.

60. The debate over these two perceptions continued through 1990 and was summed up in Bruce Parrott, "Soviet National Security under Gorbachev," in *The Soviet System in Crisis: A Reader of Western and Soviet Views*, ed. Alexander Dallin and Gail W. Lapidus (Boulder, 1991), 573.

61. Office of Technology Assessment, "Holding the Edge: Maintaining the Defense Technology Base" (April 1989), iii.

62. Ibid., 19.

63. Ibid., 19-20.

64. Ibid., 19, 22, 24.

65. Ibid., 24–25.

66. Ibid., 19, 23.

67. Ibid., 33.

68. *Washington Post*, 18 November 1988 (Thatcher), 1; *Washington Post*, 27 November 1989 ("outrunning"), 1; *Washington Post*, 27 November 1989, editorial (peace dividend), A-14.

69. Meyer interviews, 30 April and 9 May 1990.

70. Cheney, "Defense Management: A Report to the President," 1.

71. Ibid., 14-15.

72. Ibid., 26–27.

73. Office of the Secretary of Defense, "DOD Management of Technology Development—Implementation of Recommendations of Defense Management Review" (2 October 1989), 1, 4–5.

74. Ibid., 18, 23.

75. Rodney Carlisle and Joan Zenzen, *Supplying the Nuclear Arsenal: American Production Reactors, 1942–1992* (Baltimore, 1996), 200–204.

76. Staff of the DNL, "A Review of Studies Conducted from November 1989 to April 1990 on Restructuring the Navy RDT&E Community" (July 1990), 1–4.

77. Ibid., 5.

78. Ibid., 6. Schiefer spelled out his task in his interview, 1 November 1991.

79. Ibid., 8, 9.

80. Ibid., 10.

81. Schiefer interview, 1 November 1991.

82. Ibid., 16.

83. Ibid.

84. Ibid., 19–24.

85. Gerald Cann, "DOD Test and Evaluation, Research and Development Facilities Paper" (12 April 1990), 4.

86. Executive Summary of Admiral Crowe's report, as reproduced in John M. Collins, *Roles and Functions of the U.S. Combat Forces: Past, Present, and Prospects*, Congressional Research Service 93-72S (21 January 1993), 104.

87. Adam B. Siegel, *A Brave New Curriculum for a Brave New World* (Alexandria, VA, March 1991); Robert Berg and James H. Quinn, *The Navy's Operational Requirements Process: Challenges for the Future* (Alexandria, April 1990), 39. In addition to articles in defense publications like *Surface Warfare* and U.S. Naval Institute *Proceedings*, wider discussion of some of the issues took place in popular periodicals. See Michael Massing, "The Military: Conventional Warfare," *Atlantic Monthly*, January 1990, 265.

88. Defense Science Board, 1990 Summer Study, "Research and Development Strategy for the 1990s," vol. 5, Technology and Technology Transfer Task Force, 2-2.

89. Ibid., vol. 1, 1, Executive Summary, 38–42.

90. Director of Navy Laboratories, "Operation Desert Shield and Desert Storm" (31 August 1991), evaluates the support provided by RDT&E facilities to the war.

91. Ibid., 1–2.

92. Ibid.

93. Defense Base Closure and Realignment Commission, Report (1 July 1991), Ex-1, C-1, Ex-2, and Ex-3.

94. Department of the Navy Base Structure Committee, "Department of the Navy Base Closure and Realignment Recommendations, Detailed Analysis" (April 1991), 13–15, 43–48.

95. Ibid., 3–5.

96. Defense Base Closure and Realignment Commission, Report (1 July 1991), B-1, B-2.

97. See Department of the Navy Base Structure Committee, "Department of the Navy Base Closure and Realignment Recommendations, Detailed Analysis" (April 1991), Appendix F, 11–14, for the Navy and Defense recommendations; Defense Base Closure and Realignment Commission, Report (1 July 1991), 5-43 and 5-44.

98. Defense Base Closure and Realignment Commission, Report, 5-43 to 5-44, D-1 to D-2, and Ex-4.

99. Federal Advisory Commission, "Federal Advisory Commission on Consolidation and Conversion of Defense Research and Development Laboratories—Report to the Secretary of Defense" (September 1991), 3.

100. Ibid., 5.

101. Ibid., 15.

102. Ibid.

103. "Science and Technology (Techbase Strategy for the Year 2010)," NRAC 92-4 (November 1992), 5–6.

104. Ibid., 15.

105. Ibid.

106. Ibid., 21.

107. Ibid.

108. Naval Research Advisory Committee, 1993 Summer Study, 5, 41, 57.

109. "Statement of Deputy Secretary of Defense Dr. Perry Before the Senate Appropriations Committee, Subcommittee on Defense Regarding Defense Infrastructure," 20 May 1993, 8.

110. Robert D. Gourley, "Time for a Joint Ship," U.S. Naval Institute *Proceedings* (January 1994): 58.

111. Joint Directors of Laboratories, "White Paper on Tri-Service Reliance" (January 1992).

112. See Adam Siegel, *The Use of Naval Forces in the Post-Cold War Era: U.S. Navy and Marine Corps Crisis Response Activity, 1946–1990* (Alexandria, 1991).

113. A partial list of the "role and function" literature follows: Admiral William J. Crowe Jr., *Roles and Functions of the Armed Forces, A Report to the Secretary of Defense* (Washington: Office of the Chairman, Joint Chiefs of Staff, 28 September 1989); Siegel, *The Use of Naval Forces in the Post-Cold War Era*; Chairman, Joint Chiefs of Staff, *National Military Strategy of the United States* (January 1992); GAO, *United Nations: U.S. Participation in Peacekeeping Operations* (September 1992); General Colin Powell, *Roles, Missions and Functions of the Armed Forces of the United States* (December 1992); Raymond Copson, *The Use of Force in Civil Conflicts for Humanitarian Purposes: Prospects for the Post Cold War Era* (Washington: Congresssional Research Service, 2 December 1992); and Collins, *Roles and Functions of U.S. Combat Forces*. The official

Navy summary of this debate came in the joint document of CNO Admiral Frank Kelso and General C. E. Mundy, Commandant of the Marine Corps, . . . *From the Sea: Preparing the Naval Service for the 21st Century* (Washington, September 1992).

114. Kelso and Mundy, . . . *From the Sea*, reprinted in U.S. Naval Institute *Proceedings* (November 1992). The follow-on literature regarding this brief paper continued into early 1994. See, for example, General Carl Mundy, "Getting it Right '. . . From the Sea,' " U.S. Naval Institute *Proceedings* (January 1994). Articles in *Surface Warfare*, U.S. Naval Institute *Proceedings*, and other defense journals proliferated.

115. Kelso and Mundy, . . . *From the Sea*, 1–2.

116. Colonel Robert L. Venkus, *Raid on Qaddafi* (New York, 1993), 29.

117. Federal legislation that engaged Defense forces in anti-narcotics activities was the *Anti Drug Abuse Act of 1986* (PL99-570), sections 3002 and 3057, and the *National Defense Authorization Act, Fiscal Year 1989* (PL 100–456), Title XI, "Drug Interdiction and Law Enforcement Support" (29 September 1988), sections 1101–1104.

118. Michael Palmer, *On Course to Desert Storm: The United States Navy and the Persian Gulf* (Washington, 1992), xiii-xx. Les Aspin, in "The Bottom-Up Review: Forces for a New Era" (Department of Defense, 1 September 1993), classifies the Iraqi war as a Major Regional Conflict. The criteria for distinguishing between "Low Intensity" and "Major Regional" conflicts is a matter of scale; prior to Desert Storm, the Center for Naval Analyses distinguished between "Crisis Response Activity" and "A Limited War" by using the criteria of 1,000 U.S. casualties as a cutoff point. Siegel, *The Use of Naval Forces in the Post-War Era*, 2, 2n.

119. Jon P. Walman, "Expeditionary Warfare: New Emphasis for Naval Operations," *Surface Warfare*, November–December 1993.

120. Siegel, *A Brave New Curriculum for a Brave New World*; Berg and Quinn, *The Navy's Operational Requirements Process*; and James R. Blacker, et al., *Maritime Planning in a Period of Uncertainty*, CRM-91-155 (July 1991), all from Center for Naval Analyses, Alexandria, VA. The last study examines four possible "Alternative Worlds" as a basis for planning, anticipating the collapse of the USSR.

121. Kelso and Mundy, . . . *From the Sea*, 4.

122. Morton, "The U.S. Navy in Review," 118–19. This review of the whole Navy for 1992–1993 discusses the readjustment of strategic thinking, but the summary of actions at R&D facilities in the same review for the same year shows little reaction to the "littoral" emphasis stressed by the CNO. Over the same period, various officers urged one or another sort of technological response to the new strategic needs. See, for example, Robert A. Lynch, "Beyond Tomahawk," U.S. Naval Institute *Proceedings* (April 1993): 55ff. Lynch argues for improvements to the missile in response to the theme of "projecting power from the sea"; see also Floyd Kennedy, "U.S. Naval Aircraft and Weapons Developments," U.S. Naval Institute *Proceedings* (May 1993): 150, which presents the impact of . . . *From the Sea* on a variety of weapons and aircraft systems.

123. Kennedy, "U.S. Naval Aircraft and Weapons Developments," 150. Kennedy was a staff member at the Center for Naval Analyses.

124. Such experimentation by computer would be the subject of a 1994 Naval Research Advisory Committee Summer Study, under preparation at the time of the writing of this report. Colvard interview, 13 May 1994.

125. Other new approaches discussed more fully in the Cochrane article include: "Selective Upgrading," "Selective Low-Rate Procurements," "Hover," "Lean Production," "Fieldable Prototypes," "Technology Insertion," "Proof of Principle Demonstrations," and "Putting Technology on the Shelf." Charles B. Cochrane,

"DOD's New Acquisition Approach: Myth or Reality?" *Program Manager* (July-August 1992): 38–46.

126. Office of Technology Assessment, *Redesigning Defense: Planning the Transition to the Future U.S. Defense Industrial Base*, OTA-ISC-500 (Washington, July 1991); and Office of Technology Assessment, *Building Future Security: Strategies for Restructuring the Defense Technology and Industrial Base*, OTA-ISC-530 (Washington, 1992).

127. Chairman, JCS (Colin Powell), *National Military Strategy of the United States*; and Powell, *Roles, Missions, and Functions of the Armed Forces of the United States*.

128. Chairman, JCS, *National Military Strategy of the United States*, 17–23.

129. Collins, *Roles and Functions of the U.S. Combat Forces*, CRS 93-72S, 33–41, 45–60.

130. President Bill Clinton and Vice President Al Gore, study, "Technology for America's Economic Growth" (22 February 1993).

131. "John M. Deutsch, Under Secretary of Defense (Acquisition and Technology), on Technology and Industrial Base and Defense Reinvestment and Conversion Programs, Testimony Before the Subcommittee on Defense, Committee on Appropriations, U.S. Senate, 27 May 1993," 3–4.

132. Lawrence DiRita, Baker Spring, and John Luddy, "Thumbs Down to the Bottom-Up Review," *Backgrounder No. 957* (Heritage Foundation, 24 September 1993).

133. "Defense Program Exceeds Budget Target, Aspin Says," *Washington Post*, 15 September 1993.

134. Office of the Secretary of Defense, Les Aspin, "The Bottom-Up Review: Forces for a New Era" (1 September 1993).

135. For example, Norman Friedman, an author in the area of naval technology and naval history, held a seminar in the Washington area in February and March 1994 under the auspices of a private seminar firm, TTW, of Torrance, California.

136. A series of recent documents have addressed these questions. See for example, Paul G. Kaminski, "DOD's Fiscal 1997 Acquisition and Technology Program," *Defense Issues*, Vol. 11, No. 32 (Testimony, March 20, 1996). Kaminski, Under Secretary of Defense for Acquisition and Technology, has also addressed strategic issues. See, for example, "21st Century Battlefield Dominance," *Defense Issues*, Vol. 11, No. 10 (Remarks, January 16, 1996).

137. Michael A. Palmer, *Origins of the Maritime Strategy: American Naval Strategy in the First Postwar Decade* (Washington, 1988).

138. Thomas C. Hone, *Power and Change* (Washington, 1989), 131.

139. Gordon Adams, "Economic Conversion Misses the Point," *Bulletin of the Atomic Scientists* (February 1986): 24–28.

140. Seymour Melman, "Swords into Plowshares: Converting from Military to Civilian Production," *Technology Review* (January 1986): 62, 64–71.

141. Susan Dentzer,"The Peace Outbreak Sparks a War on Defense Spending: Contractors and the U.S. Economy Face a Shakeout," *U.S. News & World Report*, 4 December 1989, 51–52.

142. Bruce B. Auster, "A Healthy Military-Industrial Complex: A Diversified Defense Industry Could Help Both the Economy and the Pentagon," *U.S. News & World Report*, 12 February 1990, 42–43, 47–48.

143. *Newsweek*, 19 December 1988, 30, explores how budget cuts might fund other projects. Discussion of the "peace dividend" as a phrase was reported and evaluated in an editorial in the *Washington Post*, 27 November 1989.

144. Larry Grossman, "The Navy Comes Upon Hard Times," *Government Executive* (August 1991): 78–81.

145. C. Eugene Steuerle and Susan Wiener, *Spending the Peace Dividend: Lessons from History* (Washington, 1990); and William G. Stewart III, *From War to Peace: A History of Past Conversions; Annex B to Adjusting to the Drawdown, Report to the Defense Conversion Commission* (Washington, 1993).

146. Herbert Stein, "Remembrance of Peace Dividends Past," *American Enterprise* (March-April 1990): 18–25.

147. Robert Borosage and S. M. Miller, "Peace Dividend," *Social Policy* 20 (Spring 1990): 12–34.

148. Paul Holtgrieve and Gerald Godshaw, "The Peace Dividend," *U.S. Economic Outlook: 1989–92* (January 1990): 3.1–3.12.

149. G. Wayne Glass, *Summary of the Economic Effects of Reduced Defense Spending* (Washington, 1990).

150. George Bush, "Defense Industries: Adjusting to the End of the Cold War," in *Economic Report of the President* (Washington, 1991), 149—54.

151. Murray Weidenbaum, "The Future of the U.S. Defense Industry: Two Different Approaches," *Vital Speeches of the Day*, 15 September 1991, 722–26.

152. Murray Weidenbaum, "The Future of the U.S. Defense Industry," *Contemporary Policy Issues* (April 1992): 27–34. This article is a revised version of the speech reported in *Vital Speeches* and reflects the research for his book, *Small Wars, Big Defense* (St. Louis, MO, 1992).

153. Weidenbaum, "The Future of the U.S. Defense Industry," 27–34.

154. Bruce D. Berkowitz, "Can Defense Research Revive U.S. Industry?" *Issues in Science and Technology* 9 (Winter 1992–1993): 73–81.

155. Jurgen Brauer and John Tepper Marlin, "Converting Resources from Military to Non-Military Uses," *Journal of Economic Perspectives* 6 (Fall 1992): 145–64.

156. Ellen L. Frost, "Defense Conversion in the 1990s," *International Economic Insights* 3 (March–April 1992): 11–14.

157. Bernard L. Schwartz, *The Future of the U.S. Defense Industrial Base*, FPI Policy Briefs, (Washington, 1992), 1.

Bibliography

More than 150 other items, not included in this listing and appearing in government policy literature and academic journals, deal with the economic impact of downsizing and conversion. Those included here and in the endnotes were particularly useful in preparing this guide.

Adams, Gordon. "Economic Conversion Misses the Point." *Bulletin of the Atomic Scientists* (February 1986): 24–28.

Auster, Bruce B. "A Healthy Military-Industrial Complex: A Diversified Defense Industry Could Help Both the Economy and the Pentagon." *U.S. News & World Report*, 12 February 1990, 42–43, 47–48.

Baxter, James Phinney. *Scientists Against Time*. Boston: Little, Brown, 1946.

Bekker, Cajus. D. *Defeat at Sea*. New York: Holt, 1955.

Berg, James, and James H. Quinn. *The Navy's Operational Requirements Process: Challenges for the Future*. Alexandria, VA: Center for Naval Analyses, 1990).

Berkowitz, Bruce D. "Can Defense Research Revive U.S. Industry?" *Issues in Science and Technology* 9 (Winter 1992–1993): 73–81.

Borosage, Robert and S. M. Miller. "Peace Dividend," *Social Policy* 20 (Spring 1990): 12–34.

Brauer, Jurgen, and John Tepper Marlin. "Converting Resources from Military to Non-Military Uses," *Journal of Economic Perspectives* 6 (Fall 1992): 145–64.

Bush, George. "Defense Industries: Adjusting to the End of the Cold War." In *Economic Report of the President*. Washington: GPO, 1991, 149–54.

Bush, Vannevar. *Science: The Endless Frontier*. Washington: GPO, 1945, 1946.

______. *Modern Arms and Free Men*. New York: Simon and Schuster, 1949.

Carlisle, Rodney. *Powder and Propellants: Energetic Materials at Indian Head, Maryland, 1890–1990*. Washington: GPO, 1991.

______. *The Management of the U.S. Navy Research and Development Centers During the Cold War Era, A Survey Guide to Reports*. Washington: Naval Historical Center and Navy Laboratory/Center Coordinating Group, 1995.

______. "Where the Fleet Begins." History of David Taylor Research Center, in progress.

______, and Joan Zensen. *Supplying the Nuclear Arsenal: American Production Reactors, 1942–1992* (Baltimore: Johns Hopkins Press, 1996.

Chairman, Joint Chiefs of Staff (Colin Powell). *National Military Strategy of the United States*. Washington: Office of the Chairman of the Joint Chiefs of Staff, 1992.

Collins, John M. *Roles and Functions of the U.S. Combat Forces: Past, Present, and Prospects*. CRS 93-72S. Washington: Congressional Research Service, 1993.

Conant, James B. *Modern Science and Modern Man*. New York: Columbia University, 1952.

Copson, Raymond. *The Use of Force in Civil Conflicts for Humanitarian Purposes: Prospects for the Post Cold War Era*. Washington: Congressional Research Service, 1992.

Crowe, William J., Jr. *Roles and Functions of the Armed Forces: A Report to the Secretary of Defense*. Washington: Office of the Chairman of the Joint Chiefs of Staff, 1989.

Cuff, Robert D. *The War Industries Board: Business-Government Relations During World War I*. Baltimore: Johns Hopkins Press, 1973.

Dentzer, Susan. "The Peace Outbreak Sparks a War on Defense Spending: Contractors and the U.S. Economy Face a Shakeout." *U.S. News & World Report*, 4 December 1989, 51–52.

Devereaux, Tony. *Messenger Gods of Battle: Radio, Radar, Sonar: The Story of Electronics in War*. London: Brassey's, 1991.

Dornberger, Walter. *V-2*. London: Scientific Book Club, 1954.

Frost, Ellen L. "Defense Conversion in the 1990s." *International Economic Insights* 3 (March–April 1992): 11–14.

Gansler, Jacques S. "Industrial Contraction: Facing the Paradoxes of the Post-Cold War." *SAIS Review* 13 (Winter–Spring 1993): 105–19.

Glass, G. Wayne. *Summary of the Economic Effects of Reduced Defense Spending*. Washington: Congressional Budget Office, 1990.

Government Accounting Office. *United Nations: U.S. Participation in Peacekeeping Operations*. Washington: GAO, 1992.

Greenberg, Daniel. *The Politics of Pure Science*. New York: New American Library, 1967.

Grossman, Larry. "The Navy Comes Upon Hard Times." *Government Executive* (August 1991): 78–81.

Hannemann, Timothy. "Defense Diversification: Changing Dynamics in Aerospace/Defense." *Harvard International Review* 16 (Summer 1994): 38–39, 71–72.

Hansen, Chuck. *U.S. Nuclear Weapons: The Secret History*. Arlington, TX/New York: Aerofax/Orion Books, 1988.

Hodgson, Godfrey. *America in Our Time*. Garden City, NJ: Doubleday, 1976.

Holtgrieve, Paul, and Gerald Godshaw. "The Peace Dividend." *U.S. Economic Outlook 1989–92* (January 1990): 3.1–3.12.

Hone, Thomas C. *Power and Change: The History of the Office of the Chief of Naval Operations*. Washington: Naval Historical Center, 1989.

Kelso, Frank B., and Carl E. Mundy. *. . . From the Sea: Preparing the Naval Service for the 21st Century*. Washington: Department of the Navy/USMC, 1992.

Kistler, R. E., and R. M. Glen. *Notable Achievements of the Naval Weapons Center*. NWC TP 7088. China Lake, CA: Naval Weapons Center, 1990.

Lerner, Eric. "The Challenge of Economic Conversion." *Aerospace America* (November 1991): 26–30.

Lewis, David D. *The Fight for the Sea: The Past, Present, and Future of Submarine Warfare in the Atlantic*. New York: Collier Books, 1961.

Mandel, Robert. "The Transformation of the American Defense Industry: Corporate Perceptions and Preferences." *Armed Forces & Society* 20 (Winter 1994): 175–97.

Marolda, Edward. *The United States Navy and the Vietnam Conflict*, Vol. 2: *From Military Assistance to Combat, 1959–1965*. Washington: Naval Historical Center, 1986.

May, Ernest R. *The World War and American Isolation, 1914–1917*. Chicago: Quadrangle Books, 1966.

McDougall, Walter A. *The Heavens and the Earth: A Political History of the Space Age*. New York: Basic Books, 1985.

Melia, Tamara Moser. *Damn the Torpedoes A Short History of U.S. Naval Mine Countermeasures, 1777–1991*. Washington: Naval Historical Center, 1992.

Melman, Seymour. "Swords into Plowshares: Converting From Military to Civilian Production." *Technology Review* 89 (January 1986): 62, 64–71.

Miller, William H. "Defense Conversion—the Fourth Time Around." *Industry Week* 243 (4 April 1994): 20–22, 24, 26, 28.

Millett, Allan R., and Williamson Murray. "Innovation in the Interwar Period." Contract Report MDA903-89-K-0194. Washington: Office of Net Assessment, June 1994.

Morrocco, John D. "Pentagon to Rely on Weapons Upgrades." *Aviation Week & Space Technology* 138 (15 March 1993): 44–45, 48, 50–51.

Naval Research Advisory Committee (NRAC). "Science and Technology (Techbase Strategy for the Year 2010)." NRAC 92-4. November 1992.

______. "Defense Conversion." 1993 Summer Study. December 1993.

Norman, Colin. *The God That Limps: Science and Technology in the Eighties*. Worldwatch Institute. New York: Norton, 1981.

Office of Technology Assessment. *Redesigning Defense: Planning the Transition to the Future U.S. Defense Industrial Base*. OTA-ISC-500. Washington: OTA, 1991.

______. *Building Future Security: Strategies for Restructuring the Defense Technology and Industrial Base*. OTA-ISC-530. Washington: OTA, 1992.

______. *Defense Conversion: Redirecting R&D*. Washington: OTA, 1993.

Palmer, Michael A. *Origins of the Maritime Strategy: American Naval Strategy in the First Postwar Decade*. Washington: Naval Historical Center, 1988.

______. *On Course to Desert Storm: The United States Navy and the Persian Gulf*. Washington: Naval Historical Center, 1992.

Polmar, Norman, and Thomas B. Allen. *Rickover: Controversy and Genius—A Biography*. New York: Simon and Schuster, 1982.

Powell, Colin. *Roles, Missions, and Functions of the Armed Services of the United States*. Washington: Joint Chiefs of Staff, 1992.

Powers, Thomas. *Heisenberg's War*. New York: Knopf, 1993.

Prange, Gordon. *Miracle at Midway*. Norwalk, CT: Easton Press, 1986.

Roos, John G. "No Flight of Fancy: Sikorsky Bets on its Own Conversion Effort." *Armed Forces Journal* (May 1994): 25.

Roszak, Theodore. *The Making of a Counter-Culture*. New York: Anchor Books, 1969.

Sapolsky, Harvey M. *Science and the Navy: The History of the Office of Naval Research*. Princeton, NJ: Princeton University Press, 1990.

Schroetal, Albert H. *Military Personnel: End Strength, Separations, Transition Programs and Downsizing Strategy; Annex J to Adjusting to the Drawdown, Report of the Defense Conversion Commission*. Washington: Defense Conversion Commission, 1993.

Schwartz, Bernard L. *The Future of the U.S. Defense Industrial Base*. FPI Policy Briefs. Washington: Foreign Policy Institute, 1992.

Siegel, Adam. *The Use of Naval Forces in the Post-Cold War Era: U.S. Navy and Marine Corps Crisis Response Activity, 1946–1990*. Alexandria: Center for Naval Analyses, 1991.

______. *A Brave New Curriculum for a Brave New World*. Alexandria: Center for Naval Analyses, 1991.

Silverberg, David. "Defense Conversion: Tragedy or Farce?" *Armed Forces Journal* (May 1994): 17–22.

Spring, Baker. "Supporting the Force: The Industrial Base and Defense Conversion," *Backgrounder No. 964*. Washington: Heritage Foundation, 1993.

Stein, Herbert. "Remembrance of Peace Dividends Past." *American Enterprise* 1 (March–April 1990): 18–25.

Steuerle, C. Eugene, and Susan Wiener. *Spending the Peace Dividend: Lessons from History*. Washington: Urban Institute, 1990.

Stewart, Irvin. *Organizing Scientific Research for War: The Administrative History of the Office for Scientific Research and Development*. Boston: Little, Brown, 1948.

Stewart, William G., III. *From War to Peace: A History of Past Conversions; Annex B to Adjusting to the Drawdown, Report to the Defense Conversion Commission*. Washington: Logistics Management Institute, 1993.

Velocci, Anthony L., Jr. "Executives Blast Clinton Conversion Plan." *Aviation Week & Space Technology* 138 (24 May 1993): 27–28.

Venkus, Robert L., Colonel. *Raid on Qaddafi*. New York: St. Martin's Press, 1993.

Walker, Mark. *German National Socialism and the Quest for Nuclear Power, 1939–1949*, Cambridge and New York: Cambridge University Press, 1989.

Weidenbaum, Murray. "The Future of the U.S. Defense Industry: Two Different Approaches," *Vital Speeches of the Day* 57 (15 September 1991): 722–26.

______. "The Future of the U.S. Defense Industry." *Contemporary Policy Issues* 10 (April 1992): 27–34.

______. *Small Wars, Big Defense: Living in a World of Lower Tensions*. St. Louis, MO: Center for the Study of American Business, 1992.

Weir, Gary E. *Forged in War: The Naval-Industrial Complex and American Submarine Construction, 1940–1961*. Washington: Naval Historical Center, 1993.

Index